THE ART OF RECONNAISSANCE

THE ART OF RECONNAISSANCE:

With Analytic Annotations

by

Henry Prunckun

Bibliologica Press

The Art of Reconnaissance:
With Analytic Annotations
by Henry Prunckun

Text by Sir David Henderson, April 1914
Introduction, annotations, and index copyright © 2019 by Henry Prunckun

ISBN 978-0-6485093-0-1

NATIONAL
LIBRARY
OF AUSTRALIA

A catalogue record for this
book is available from the
National Library of Australia

Published by Bibliologica Press
P.O. Box 656
Unley, South Australia, 5061
Australia

TABLE OF CONTENTS

ACKNOWLEDGEMENT

Major-General Sir David Henderson's third edition was originally published by John Murray, London, 1914—later acquired by Hodder & Stoughton, London—and is reproduced here with the publisher's acknowledgement.

First Edition	July 1907
Second Edition	October 1908
Reprinted	April 1911
Third Edition	May 1914
Reprinted	January 1915
Reprinted	November 1915
Reprinted	August 1916

INTRODUCTION

In 1907, then-Major-General[1] Sir David Henderson, K.C.B.,[2] D.S.O.,[3] wrote the watershed military text on reconnaissance. Titled, *The Art of Reconnaissance*, he scribed what was the equivalent to Sun Tzu's *The Art of War*, but for the topic of reconnaissance. Yet, it never gained the circulation that the Chinese strategist enjoyed. Maybe this was because Henderson's text was more practically orientated, and hence operational in its application. Sun Tzu, in contrast, considered strategic issues that would later be applied to other forms of conflict (e.g., business) and was able to be transported from the fifth century B.C., when he wrote, to modern times.

Nevertheless, Henderson's treatment of the art of reconnaissance deserves to be resurrected because the need for reconnaissance has never abated. Militaries still have reconnaissance units and this activity has been used in every major conflict since Henderson fought in the Second Boer War (1899–1902). In fact, there are historical references of reconnaissance being used by the armies during the American Civil War, as well as in ancient times.

---

1. In 1917 to become Lieutenant-General and in 1919 to be award the Royal Victorian Order (KCVO), which is an order of knighthood under royal patronage for distinguished personal service to the monarch.

2. Knight Commander of the Bath is a British military award.

3. Distinguished Service Order is another prestigious military decoration awarded by the United Kingdom Government.

Henderson served as Lord Kitchener's director of military intelligence from 1900 to 1902 because of his new, more scientific approach to intelligence gathering. His innovative way of viewing military intelligence saw him publish two manuals—*Field Intelligence: Its Principles and Practice* (1904) and *The Art of Reconnaissance* (1907). The latter is the subject of this analytically annotated edition.

Lt-Gen Sir David Henderson, Royal Flying Corps, circa 1916–1918. Courtesy of the British Government.

Leading up to the Great War,[4] *The Art of Reconnaissance* found a large readership as evidenced by its reprint demand—a second edition in 1908, which was again reprinted in April 1911. The third edition, which is the subject of this critique, was published in May 1914 and reprinted three times; in January 1915, November 1915, and August 1916.[5]

Although Henderson's text is grounded in a form of warfare that is no longer practiced—horse mounted operations—the principles and practices of the art can be applied in the modern context. *The Art of Reconnaissance* shows that not only is the art

---

4. Before the Second World War, the global hostilities of 1914–1918 were commonly referred to as the *Great War*, or just the *World War*. Paul Fussell, *The Great War and Modern Memory* (Oxford: Oxford University press, 1975).

5. The version presented here was published by John Murray, Albemarle Street, W., London, 1916.

that Henderson espoused over a century ago still relevant today, but his scientific way of thinking has been—intentionally or unconsciously—incorporated into different aspects of present-day intelligence gathering.

I hope my analysis of Henderson's somewhat forgotten text will provide insights for scholars and practitioners alike; and in doing so, hopefully provide an interesting read for those who are simply interested in military, national security, law enforcement and private sector intelligence work.

In presenting Henderson's book, the reader will note a few peculiarities with his writing. Given the era when Henderson wrote, the style of writing, language, and expression are different to that which is popular at the time of this writing. Generally, authors have moved from a style of writing that was characterized by long paragraphs, compound or complex sentences, verbosity, and convoluted language. We now refer to this simpler style as *plain language* or *plain English*.[6] The other aspect that readers will note is regarding punctuation. Grammarians will no doubt see issues in the way the original publisher edited, or failed to edit, Henderson's manuscript. Nonetheless, the text is presented here as it appeared in the 1916 printing of the third edition.

Finally, I will point out how I have annotated the text. In the third edition of *The Art of Reconnaissance*, Henderson used footnotes to reference some parts of his writing. Although today this brief style of footnoting finds itself wanting in terms of providing adequate biographic information. Regardless, it was accepted at the time he wrote because, perhaps, it did not burden

---

6. For instance, see Martin Cutts, *Oxford Guide to Plain English, Fourth Edition* (Oxford: Oxford University Press, 2013).

the reader with detail; after all, these were action-orientated military personnel who read his book. Today, this brief style of footnoting might raise eyebrows in scholarly circles, especially given the number of quotes Henderson relies on without attribution to the source.

I have reproduced Henderson's footnotes as they appear in the original text, but as endnotes using alphabetical notation to distinguish them from my analysis. Henderson's original line maps have been reproduced as they appeared in the third edition.

My analytic annotations of key ideas, scholarly thoughts and reflections are contained in numeric footnotes[7] with the goal to generate a new understanding of this area of study; one that is based on research and grounded in intelligence theory.

Henry (Hank) Prunckun, BS, MSocSc, MPhil, PhD
Research Criminologist
Australian Graduate School of Policing and Security
Charles Sturt University
Sydney

July 2019

---

7. I have used the Chicago referencing style because it has wide use in the humanities, including history.

PREFACE TO THE THIRD EDITION

In the seven years which have elapsed since this book was first published, the power of flight has been added to the resources of mankind. Whatever may be the influence of the art of flying on the future developments of civilization, this influence is, as yet, apparent only in one direction, that of warfare, and in its application to warfare the new art has already affected reconnaissance.

A book on reconnaissance, in which the possibilities of aerial scouting are not considered, must be classed as obsolete; and yet, in the new field, experience is so limited, and progress is so rapid, that any dogmatic pronouncement on military aeronautics would, at this stage, have no permanence and little value. I have therefore in this new edition done no more than add a chapter on aerial reconnaissance, in which will be found some indication of the results which have already been produced by the introduction of a third dimension into the problems of war.[8]

April 1914
D.H.

---

8. Henderson showed astute insight into what aviation would mean for the military. His prophetic words have come true and his comments on the rapid progress are modestly understated.

PREFACE

This work is intended and hence its pretentious title to present a view of the subject reconnaissance as a whole, in the hope of assisting those whose duty or ambition it is it may be to prepare themselves to undertake the pursuit of information in war. There are two points which may seem to call for some explanation. If anyone should remark on the incompleteness of the work and should complain that in discussing the details of reconnaissance, I have left a great deal unsaid, I would reply that I have endeavored to consider only those details which seem to contain the germ of some principle of more or less general application. If, on the other hand come up exception should be taken to the temerity of a foot-soldier in surveying, and perhaps overstepping the debatable ground which lies between the provinces of reconnaissance and cavalry tactics, my defence is that I went in search of knowledge, and that such a quest is a reasonable excuse for trespassing.

[July 1907]
[D.H.][9]

---

9. Although unsigned in the original version, this is Henderson's preface to the first edition that was published in July 1907.

CHAPTER I

PRINCIPLES AND METHODS

The acquisition of information about the enemy has always been considered one of the most important elements of success in war. A commander without information is like a man blindfolded; he knows neither where to strike nor from what quarter to expect attack; he is unable to make a plan for himself, or to guard against the plan of his enemy. It is therefore the aim of every general to gain information about his adversary, and to deny to his adversary information about himself.[10] He who succeeds in this object has gained a preliminary advantage. The value of the advantage cannot be assessed accurately, even for historical cases; the difficulty and complexity of the art of war is due, in great measure, to the number and the varying importance of the different factors which together produce success. The strategical or tactical skill of commanders, the valour, discipline, and endurance of the troops, organisation, numbers, weapons, luck: these also must be considered by the student who searches for the secret of victory. Their relative importance can only be a matter of opinion. The popular method of ascribing success to some particular advantage—to the strategy of Turenne or the tactics of Condé, to the ardour of the French or the doggedness of

---

10. Although Henderson's two opening sentences appear to form a simple truism, this is a thoughtful opening to the book because it explains what intelligence can do, and why. It is not only military commanders who need an intelligence capability, but others who engage in conflict—including *business competition*—that need to practice intelligence.

the British, to the Roman sword or the German needle-gun[11]—
shows ignorance of war. The most that can be said is that in
certain campaigns or battles one advantage has overshadowed the
rest. The strategy of Ulm,[12] the tactics of Salamanca,[13] the
discipline of Dettingen,[14] the organisation of Sédan,[15] are
generally accepted as decisive influences. Information cannot be
classed with these factors, because its influence is indirect, while
theirs is direct. Its importance lies in the fact that it is essential to
the success of both strategy and tactics, in fact to the making of
plans. If he has superior information, a mediocre commander
may be able to meet a master of strategy on equal terms; to a
skillful leader, with good information, anything is possible. The

---

11. The name *needle-gun* was derived from the German
Zündnadelgewehr, which referred to the weapon's needle-like firing pin.
Unlike other long-arms of the 1800s that were muzzle-loaded, the gun's
firing pin struck a percussion cap that was at the base of a paper cartridge
that also contained the powder and bullet. It is credited to be the invention
of Johann Nikolaus von Dreyse as the first breach loading rifle with a bolt
action. The combination of a self-contained cartridge (i.e., percussion cap,
power, and bullet) brought the Prussian army a great advantage. The
needle-gun was put into service by the Prussians in 1841.

12. Refers to the Battle of Ulm, which took place between 25 September
and 20 October 1805, between the Grand Army under the command of
Napoléon and the Austrian Army, commanded by Baron Karl Mack von
Leiberich.

13. Refers to the Battle of Salamanca, which involved an Anglo-
Portuguese army under the command of the Duke of Wellington, which
defeated French forces south of Salamanca, Spain on 22 July 1812 during
the Peninsular War.

14. The Battle of Dettingen on 27 June 1743 took place at Dettingen
Germany during the War of the Austrian Succession.

15. The Battle of Sédan in north eastern France was fought between 1
and 2 September 1870 during the Franco-Prussian War.

fateful combinations of Napoleon[16] and the astonishing enterprises of Lord Peterborough[17] were alike based on information; trustworthy information for themselves, and no information, or that which was false, for their adversaries.

The acquisition of accurate information is one of the most difficult tasks of a commander in the field. The numbers, the dispositions, the movements of the enemy are veiled in an obscurity which has been aptly named the fog of war—an obscurity which the opposing general will endeavour, by every artifice, to deepen. The value of information is thus enhanced by the difficulty of obtaining it; crumbs are welcome to a starving man. Therefore, it is that as soon as armies take the field, the pursuit of information becomes a serious, perhaps the paramount consideration in the mind of a commander. He feels that, unless he can in some way pierce the fog which surrounds his adversary, he will be unable to devise a scheme either to compass that adversary's over-throw, or to ensure his own safety. Yet so indirect is the influence of information, so much is its value overshadowed by the evident effect of strategical and tactical combinations, that students of war sometimes forget that such combinations are entirely dependent on information; and consider historical problems without regard to the amount of information available on either side, still less to the methods by which the information was acquired. Successful generals, too, are occasionally inclined to minimise the assistance they have received from good information, fearing that if they admit advantage in this respect, the influence of their own skill, or

---

16. Napoléon Bonaparte (1769–1821), not to be confused with his nephew, Napoléon III (born Charles-Louis Napoléon Bonaparte, 1808–1873).

17. Lord Peterborough, Charles Mordaunt, 3rd Earl of Peterborough and 1st Earl of Monmouth.

intuition, or genius, on the result, may be discounted; it is only the unsuccessful general who emphasises the importance of the information which he did not get, and nobody pays much attention to the excuses of the man who failed.

In peace time, therefore, the importance of information has in the past fallen rather into the background, and the study of the means by which it may be obtained has been neglected.[18] Of late years, however, the methods of military study have changed; the development of the system of the General Staff, in continental countries, has substituted for the individual student the expert and well-equipped council, engaged continuously on the solution of military problems. History has been ransacked in the search for the principles of victory; speculation on the possibilities of the future has been pushed to its furthest limits. Naturally, one of the first historical discoveries was the proof of the value of information in the past;[19] one of the first speculations was as to the method by which it could best be acquired in the future. Reconnaissance immediately took its proper place as one of the

---

18. This appears to be a timeless issue. Political leaders who oversee the budgets of intelligence agencies need to heed the words of former CIA director Robert Gates: "The nation is at peace because we in intelligence are consistently at war." Gates, cited in Charles Lathrop, *The Literary Spy: The Ultimate Source for Quotations on Espionage and Intelligence* (New Haven, CT: Yale University Press, 2004), 205.

19. Known as *basic intelligence*. "However, the term basic intelligence is a bit of a misnomer. The term infers that somehow it is elementary or simple, but it is neither of these things. Basic intelligence is concerned with analysing historical topics. The purpose is to provide information that can be used for a variety of research projects, as well as for operational reasons. An example of the latter is where an operative or agent is developing a "cover" or "legend" and needs facts about what a place looked like at a particular time." Hank Prunckun, *Methods of Inquiry for Intelligence Analysis, Third Edition* (Lanham, MD: Rowman & Littlefield, 2019), 15.

most important branches of the military art, with the result that the organisation and training of cavalry are now designed to fit them primarily for this duty; and even in the slow-moving infantry, ambitions towards the pursuit of information have developed.

An example of the effect which the possession or the lack of good information may exercise on an operation of war may be found in a comparison of two battles—Friedland and Spicheren—between which, in the opening situation, there is a curious similarity. In each case, the French troops were disposed in a wide arc in the vicinity of a river crossing by which it was expected that the enemy would advance. There were five French *corps d'armée*[20] and some 11,000 cavalry[21] (about 80,000, all told) within 25 miles of Friedland;[22] there were four *corps d'armée* and four cavalry divisions (130,000 men) within 28 miles of Spicheren.[23] In each case a single corps was advanced to the crossing itself, and had to bear the brunt of the attack. At the close of the battle of Friedland, the whole of Napoleon's 80,000 men were present on the field, and the enemy had been over-whelmed; at the close of the battle of Spicheren, Frossard's[24] solitary corps was retiring sullenly from a position

---

20. In English, *army corps*.

21. In simple terms, cavalry fought on horses with swords and lances, whereas the likes of the Australian light horse regiments (fashioned after the dragoons) were mounted infantry who used their horses for mobility but dismounted to fight on foot.

22. The Battle of Friedland took place on 14 June 1807 and was considered by military historians as one of the major engagements of the Napoleonic Wars.

23. The Battle of Spicheren was fought on 6 August 1870 during the Franco-Prussian War.

24. French general, Charles Auguste Frossard (1807–1875).

which it had held, unsupported, all day, and, save for one dragoon brigade, no other French troops had been in action. Making every allowance for the genius of Napoleon I, and for the mediocrity in a military sense of Napoleon III, and of his lieutenant, Bazaine;[25] admitting skill and good fortune in one case and faults and bad luck in the other; waiving even the consideration that Lannes was outnumbered from the first, while Frossard held the advantage of numbers till late in the day; it is yet evident that some other influence must have been at work to produce such divergent results from premises so similar. Study of the campaigns makes the matter clear at once. The controlling influence was information: Napoleon I had good information; Napoleon III had none. Friedland was fought on 14th June 1807. On the 11th June, "Napoleon left upon the Alle the rest of his cavalry, composed of chasseurs,[26] hussars,[27] and dragoons,[28] for the purpose of beating the banks of that river and closely pursuing the enemy."[A] On 13th June, "The indications collected respecting the march of General Benningsen were as uncertain as the plans of that general. On the one hand, the light cavalry had followed the main body of the Prussian army along the Alle, had seen it between Bartenstein and Schippenbell; on the other, it had been imagined that detachments of the enemy had been

---

25. François Achille Bazaine served as an officer in the French army from 1831 to 1873.

26. Light infantry who were trained in marksmanship and could execute quick manoeuvres.

27. Used in several roles, such as reconnaissance, harassing enemy skirmish lines, overrunning artillery positions, and pursuing retreating troops.

28. Mounted infantry.

perceived going to Königsberg."[B, 29] By the evening of the 13th, "the reconnaissances of the day left no further doubt that General Benningsen had descended the Alle, and appeared to be taking the road to Friedland, either *History of the Consulate and the Empire*, Thiers, to continue his march along the Alle, or to leave there the banks of that river, in order to gain Königsberg. ... Napoleon hesitated not a moment longer.

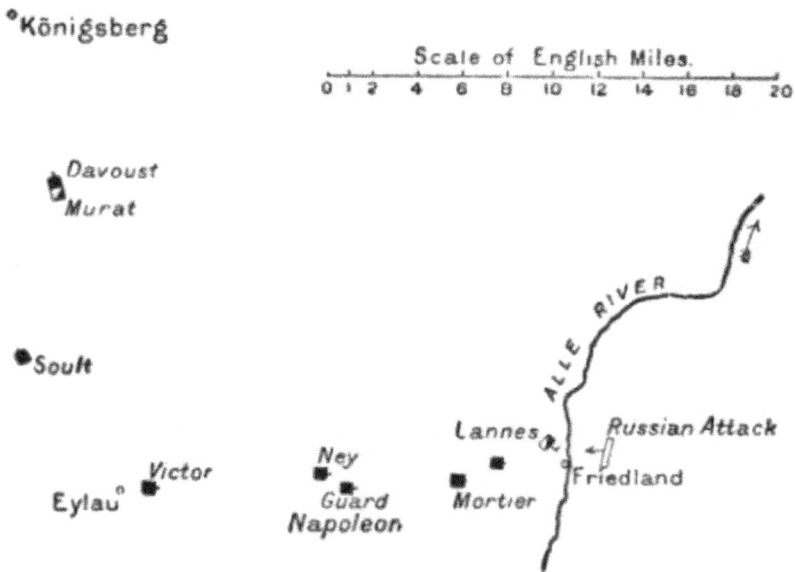

FRIEDLAND.

French positions on morning of 14th June 1807.

---

29. Louis Adolphe Thiers, translated by D. Forbes Campbell and John Stebbing, *History of the Consulate and the Empire of France Under Napoleon* (London: Chatto & Wingus, 1893–1894, twelve volumes).

He dispatched towards Lannes and Mortier all that part of the cavalry which had not followed Murat,[30] and gave the command of it to General Grouchy.[31] He enjoined Lannes and Mortier to proceed to Friedland. … He ordered Ney and Victor to follow Lannes. … He then marched off his guard."[C] Here there is information, and evidence of the search for it. Very different was the case in 1870. After 24th July, the French had 76 squadrons available for reconnaissance duty. "The reconnaissances effected, however, by this imposing army of horsemen, although numerous enough, were carried out without energy or enterprise. They were undertaken, as a rule, in conjunction with infantry and artillery, and the combined detachments were unable either to move rapidly or to conceal their approach. The same villages were visited at the same hours, and by the same roads; and the enemy's patrols, made aware by the country people of the time and route at and by which the French were to be expected, easily avoided them.

No attempts were made by small parties to pierce the hostile lines. Nor was this powerful force of 8,000 men able to prevent the approach, at all hours and at every point, of the Prussian scouts. At French head-quarters, therefore, the most absolute ignorance pre-vailed as to the points of concentration and the whereabouts of the Prussian Army Corps. No single item of intelligence was permitted to filter through the line of outposts on the Saar.[32] The German press maintained a discreet silence as to military movements, and Napoleon's staff had to rely for their information on the columns of the English newspapers, or the reports of double-dealing spies. The leaders of the French *corps*

---

30. Joachim-Napoléon Murat (1767–1815) served as Marshal of France and Admiral of France under the Napoléon's reign.

31. General Emmanuel de Grouchy, 1779–1815.

32. A region in western Germany.

d'armée along the frontier, who had each of them four regiments of cavalry at least attached to their command, appeared to have been fettered in their employment by instructions from headquarters. Thus, General Frossard, commanding the Second Corps, the advanced guard of the Metz force, was ordered by Marshal Bazaine, in temporary charge of the left wing, to reconnoitre from St. Avold, but only as far as the frontier, and without compromising any important detachment. ...

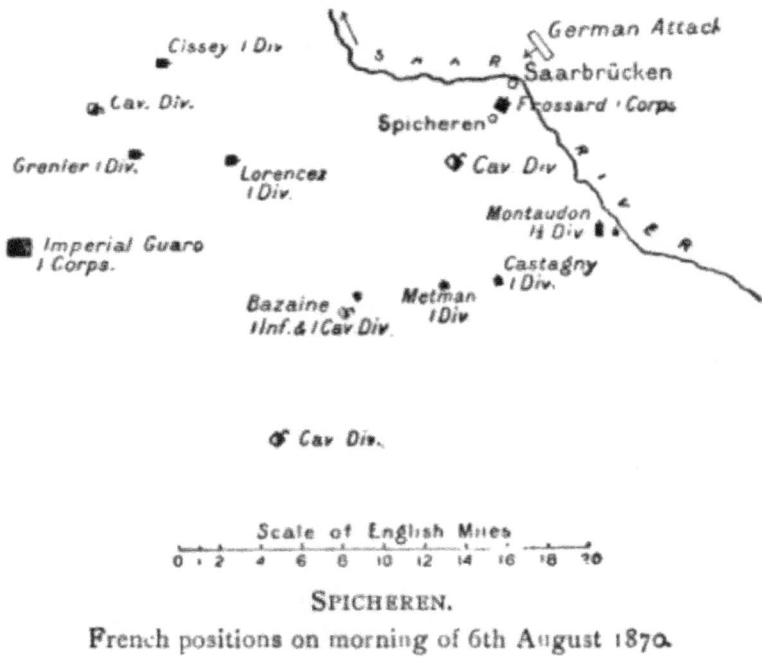

SPICHEREN.

French positions on morning of 6th August 1870.

The business of reconnaissance, of covering the front, so efficiently performed in the wars of Napoleon, was a lost art in

France."[D, 33]   On the day of battle itself, 6th August, the information was no better.

"On the evening of the 5th, he (Bazaine) had been placed in command of the French left wing; but, in consequence of the failure of the cavalry to procure information, he was without knowledge of the enemy's dispositions, and unable to conjecture where the attack that was believed imminent would fall. ... The French . . . had neither patrols nor scouts. They were ignorant of the whereabouts of the hostile masses, unable to forecast with any degree of precision the point where the enemy would strive to break their line—Saarbrucken, Saarlouis, or Saargemund.[34] . . . When the cannon was heard from the direction of Spicheren, and even when reports came in that Frossard was heavily engaged, doubts must have arisen whether this was not a feigned attack, and apprehensions have been excited that the real blow was about to fall elsewhere."[E] Nor did Frossard himself give much aid in clearing up the situation. At five o'clock in the afternoon Bazaine telegraphed to him: "Donnezmoi des nouvelles,[35] pour me tranquilliser."

It may safely be said that there is no general who ever fought an action who has not echoed Bazaine's cry. A commander's thirst for news, for information, is insatiable.[36] It is by efficient reconnaissance that this thirst can best be allayed.

---

33. Major G.F.R. Henderson, *The Battle of Spicheren, August 6th, 1870 and the Events that Proceeded it. A Study in the Practical Tactic and War Training* (London: Gale & Polden Ltd., 1891).

34. Three closely situated towns in western Germany, all of which are southeast of Luxemburg.

35. *Donnez-moi des nouvelles* translates to "give me news."

36. This axiom has been developed to the point where commanders now have real-time reporting from forward observers, drone surveillance, and satellite reconnaissance.

Reconnaissance is usually understood to mean the acquisition of information about an enemy, or about a country, by personal observation; and in time of war it is confined to the operations, to that end, of combatants in uniform. It is by means chiefly of reconnaissance that a commander in, the field endeavours to ascertain the visible numbers, dispositions, and movements of his enemy, and to obtain such detailed information about the theatre of operations as may be necessary to supplement the maps at his disposal. The other means of gaining the required knowledge are the collection of second-hand information from the statements of prisoners, inhabitants, deserters, or others, or from documents, and the employment of secret service. Reconnaissance is the most important of all these means; for, in the first place, the security of an army depends always in great degree on the vigilance of its outposts and patrols; and secondly, the success of its operations is contingent on knowledge of the enemy's immediate dispositions.

Reconnaissance in an efficient army in the field should be unceasing, and there is no limit to the proportion of the whole force which may be employed at any one time on this duty. The exigencies of the moment dictate the number of men to be detailed for reconnaissance duty; at one time a few patrols may be sufficient, at another a cavalry division may be necessary; it is conceivable that even the whole force might be justifiably committed to action with the principal object of clearing up an unbearably obscure situation.

In order to consider clearly the principles on which reconnaissance should be conducted, it is necessary to distinguish between, and to define, the different systems which may be used. There are three types of reconnaissance of the enemy, which differ essentially from each other in the methods by which the same object—information—is obtained; these may be called protective, contact, and independent reconnaissance.[37] The first, protective, is that which is confined to ensuring the absence of the enemy, or to obtaining just sufficient warning of his proximity, and to the prevention of the enemy's efforts to obtain information. It is carried out by the outposts of a force, by patrols of limited range, by detached forces, by flank and rear guards, and in some cases by advanced guards. Reconnaissance of this kind is used for purposes of security or of protection, or of both; a force detached to screen from the enemy the dispositions or movements of an army, must provide for its own security, as well as for the protection of the army. In either case, however, the enemy is sought for only within a certain defined distance, which is calculated so as to be sufficient to preclude the possibility of surprise.

The second, contact reconnaissance, is that employed by large bodies, which seek out the enemy and are prepared to fight, if necessary, for the information they desire. This method is employed in the case of a reconnaissance in force; it is usually followed by a large cavalry force when detailed for reconnaissance, and in many cases by advanced guards.

---

37. These three types of reconnaissance have remained the same, though independent reconnaissance is also termed "autonomous" reconnaissance. In addition, there is a fourth type—special reconnaissance. "This usually involves deep penetration into an area controlled by an [enemy] or into an area that will be in some way unsympathetic to the presence of the reconnaissance unit. ... The mission is based on covert insertion, maneuvering and extraction ... and is sometimes referred to as *black recon.*" See Henry Prunckun, *How to Undertake Surveillance and Reconnaissance: From a Civilian and Military Perspective* (South Yorkshire: Pen & Sword Books Ltd, 2015), 27–30.

The third, the independent system,[38] is that followed by patrols and scouting parties, whose discretion is wide and range unlimited, which seek to obtain their information without being observed, and are not intended to fight but to elude the enemy. The reconnaissance of ground for topographical purposes is not a duty for which any special method is required. If the ground be occupied by the enemy, then its configuration would naturally be considered in any reconnaissance of that enemy; if the district to be surveyed be clear of the enemy, then the interest centres in the information itself, and not in the more difficult part of reconnaissance—the consideration of the means by which the information is to be obtained.

Of late years the possibilities of reconnaissance have been greatly enlarged by the invention of aeroplanes and dirigible airships,[39] and in the wars of the future there can be no doubt that the use of aircraft will make the acquisition of information, both strategical and tactical, more certain and more easy than in the past.[40] It may be frankly conceded that aerial reconnaissance is, even now in its early stages of development, the method by which information of any considerable forces of the enemy can be obtained most rapidly, accurately, and completely.[41] It is, however, at present subject to certain very definite limitations; in fog, storms, or darkness, aerial observers find it difficult, and may find it impossible, to gain information. Therefore, although observation from aircraft is already an invaluable and indispensable method of reconnaissance, it cannot yet replace

---

38. More generally referred to as *autonomous reconnaissance*.

39. In contrast to a balloon that drifts in currents of air, a dirigible is able to be steered by a pilot.

40. A statement taken for granted now, but astutely prophetic at the time.

41. And, this has held true ever since.

completely any of the methods which have been employed hitherto; as a rule, infinitely more information will be available; in the exception, the old methods must still be relied on.[42]

In every branch of reconnaissance aerial observation will be of assistance. Strategically the masses of the enemy may be sought out and marked down even before their deployment is complete; the task of contact cavalry and of independent patrols is thus rendered infinitely less arduous and costly. The approach march of columns may be watched, the occupation of positions observed, and tactical combinations reported; and so long as information from this source is sufficient, the cavalry can be spared and nursed, held in readiness to act as a reconnoitring force, should the elements prevail against the aircraft, or as a fighting force, should opportunity offer. An army in the field ought to be organised in such a way as to be able to carry out all systems of reconnaissance simultaneously and continuously. Protective reconnaissance of some kind is a duty in which all combatant forces must be prepared to take part. Outposts at the halt, advanced or rear guards on the march, are indispensable, in order that warning of the enemy's proximity may be obtained. With small forces, the service of protection may well be carried out by infantry, for the screen need be only at such distance as will prevent actual surprise; but for large forces, it is necessary that sufficient warning should be given to enable the commander to make dispositions for action; and the amount of time required

---

42. I would argue that Henderson foreshadowed the advent of technological developments that would eventually overcome these impediments. Although it was beyond his ability to speculate how and in what way such technology would evolve, his preface that "at present [it is] subject to certain very definite limitations" anticipates that engineers would eventually develop solutions. Take for instance the development of infrared photography, cloud penetrating radar, night vision viewing devices, and thermal imaging, to mention a few of the solutions.

for this increases with the strength of the force. It is customary, therefore, to allot to divisions and army corps or armies, bodies of mounted troops, the special function of which is protective reconnaissance: and it is to these troops that the protection of large forces should be mainly entrusted.

The force provided for contact reconnaissance is the "independent" cavalry, which is organised in divisions or brigades, and is properly under the immediate orders of the commander-in-chief. Its duty is to gain and keep touch with those forces of the enemy which are in the immediate theatre of operations. To accomplish this task, the cavalry must be prepared to fight; the recurrence of such opportunities as were offered to the German cavalry in 1870, can hardly be hoped for. The opposing cavalry must be dealt with, for mere contact with the hostile mounted men is not sufficient to constitute what is known as touch with the enemy's forces. It is necessary to break or drive in the enemy's advanced troops, both tactical and protective to locate and if possible, identify the formed bodies of his main forces, to discover his movements. This demands resolute and combined action and may entail concentration at any moment and in any direction to operate tactically against the hostile mounted troops. It is clear, therefore, that the initial dispositions and the preliminary operations of the independent cavalry, should be based as much on tactical as on reconnaissance considerations, and also that cavalry employed on contact reconnaissance cannot, without losing its freedom of manoeuvre, simultaneously perform protective duties.

The distinction between the forces allotted to contact and to protective duties has been to some extent obscured by the indefinite names which have been applied to them. The force which is usually called "independent" cavalry is by General von

Pelet-Narbonne, in his book *Cavalry on Service*,"F, 43 named "army" cavalry. Neither name is quite satisfactory, but the former is preferable as indicating that a force of this kind is under the orders only of the commander-in-chief, the generalissimo, and is independent of all subordinate commanders. If several armies are operating in conjunction, the term "army" cavalry becomes a misnomer, for the independent cavalry has not necessarily any connection with a particular army but is the strategical weapon of the chief of all the armies. The modern theory is that, under such circumstances, the whole of the cavalry available for contact, or offensive reconnaissance should be massed, and should operate under the direct instructions of the supreme commander, for the general benefit of all the armies, and that it is a waste of power to divide the cavalry and allow each army commander to carry out his own contact reconnaissance. Colonel Cherfils[44] in his *Essai sur l'emploi de la Cavalerie*,[45] refers to this offensive cavalry as "independent cavalry," "the cavalry corps," or "the cavalry of the generalissimo"; and of these, the term "independent," being in common use, is perhaps the most suitable for general purposes; but in considering details of reconnaissance, it is more convenient to refer to it as "contact" cavalry.

---

43. Gerhard von Pelet-Narbonne, translated by Major D'Arcy Legard, *Cavalry on Service: Illustrated by the Advance of the German Cavalry Across the Mosel in 1870* (London: Hugh Rees, 1906).

44. General Pierre Joseph Maxime Cherfils (1849–1933).

45. The full citation (in French) is, *Essai Sur L'emploi De La Cavalerie: Leçons Vécues De La Guerre De 1870 Et Faites En 1895 A L'école Supérieure De Guerre.* An approximate English translation is: "Analysis on the Employment of the Cavalry: Lessons Learned from the War of 1870 and Made in 1895 at the Superior School of War."

Protective cavalry is usually designated by the name of the formation to which it is attached, as "army," "army corps," or "divisional" cavalry. This system of nomenclature is, again, sufficient for purposes of organisation, but it is cumbrous and indefinite. The term "protective" cavalry, which is now used in British regulations, is both clear and comprehensive.

In order to supplement both contact and protective reconnaissance, the system of independent reconnaissance is of the greatest value. The troops allotted to contact reconnaissance may be for a long time engaged in tactical operations against the enemy's cavalry; the protective squadrons may see no enemy; the only real source of information by reconnaissance is then the independent patrol or scouting party. These are sent out on the initiative of commanders by whom information is required and may be composed of men from any branch of the service. Their one aim should be the acquisition of information, and every detail of their procedure should be adapted to this end. The necessity for a clear understanding of the distinction between the different types of reconnaissance is well brought out by General Pelet-Narbonne in his account of the advance of the German Cavalry across the Mosel[46] in August 1870. At the beginning of the war the German leaders, except perhaps Prince Frederick Charles,[47] had no real appreciation of the functions of cavalry in the matter of reconnaissance; General von Steinmetz,[48] for example, asserted that the cavalry should be kept in rear. Thus,

---

46. The Moselle (in French, and Mosel in German) is a river that flows through France, Luxembourg, and Germany, becoming the left tributary of the Rhine.

47. At the start of the Franco-Prussian War, Prince Friedrich Karl Nicolaus of Prussia (1828–1885) was put in command of the Second Army and was distinguished at the Battle of Spicheren.

48. Karl Friedrich von Steinmetz (1796–1877).

on 7th August, the day after the battle of Spicheren, the three
available cavalry divisions of the First and Second Armies
remained practically stationary, although the tracking of a
defeated enemy is both an important duty and the supreme
opportunity of the troops detailed for contact reconnaissance. In
this case, the only effort to gain touch was the despatch of
several officers' patrols, of which the most successful was one
sent out by a regiment of divisional cavalry. Some of these
patrols came into contact with various bodies of French troops,
but being unsupported, were unable to maintain touch. In spite
of the extraordinary inaction of the French cavalry, it was not
until the 10th August that sufficient information to warrant the
advance of the German armies was obtained, and this time it was
an officers' patrol from the headquarters of an army corps which
made contact, although five divisions of cavalry were available
for the task. There were various reasons for this failure. The
country was difficult, the weather was bad, and the training of
the German cavalry in combined reconnaissance was insufficient.
Nevertheless, the principal fault was the misunderstanding of the
duties and handling of cavalry. The cavalry divisions, instead of
being controlled by royal or army headquarters, or united under
the leadership of a single responsible commander, were attached
to different army corps, and the army corps' commanders either
kept them in rear or employed them on protective
reconnaissance. General Pelet-Narbonne puts the case clearly:
"The deployment of the armies advancing from their detraining
points was to be protected and concealed, and information was to
be obtained about the distribution, movements, and intentions of
the enemy's forces. These duties, which fall to the cavalry, were
of both an offensive and defensive nature, and must be separately
carried out to attain a successful result—i.e., the duty of
observation must be fundamentally separate from that of
protection. Only the first of these duties properly belongs to the
army cavalry, in this instance the cavalry divisions; the second

belongs to the divisional cavalry, if these, as in this case, are available in such quantities, otherwise they would be strengthened for the purpose by the cavalry divisions."

There may be circumstances in which it is advisable to employ the independent cavalry on protective reconnaissance, and an instance in point is the action of the French cavalry under Murat in the campaign of Ulm in 1805. In this case, it was essential that the movements of the French armies should be concealed; and practically all the available cavalry was extended as a screen, which effectually veiled Napoleon's strategic design. Protection was, in fact, more important than information, for information was available from other sources. It was not until the great strategic combination was nearly affected that Murat changed his method and assumed a swift offensive. The contact which he then established was in a great degree responsible for the startling success of the campaign.

A somewhat similar plan was adopted by the Archduke Albert of Austria,[49] in the campaign of Custozza in 1866.[50] The Austrian forces, greatly outnumbered by the Italians, were on the defensive; but the Austrian commander-in-chief considered that, as his adversary's forces were divided, it was possible that an opportunity for offensive action might be offered him, provided he was able to keep the Italians in ignorance of his dispositions.

---

49. i.e., Archduke Albrecht Friedrich Rudolf Dominik of Austria (1817–1895).

50. The battle took place on 24 June 1866 during the Third Italian War of Independence. The war was between the Kingdom of Italy and the Austrian Empire and was fought between June and August 1866. The conflict took place at the same time as the Austro-Prussian War and resulted in Austria's defeat. Austria turned over political control of what was the Venetia region, which contributed in part to the Italian unification process, which eventually brought together the different Italian states.

For this reason, he employed the whole of his available cavalry—two brigades—on protective reconnaissance, and supported these outposts with infantry, in order to make a continuous screen, impenetrable unless attacked in force. The Archduke had, by information from other sources, satisfied himself as to the general lines on which the Italians would invade Austrian territory; for the moment, therefore, protection was of more importance than information. But as soon as the Italians had committed themselves to a definite advance, the Austrian cavalry dropped the protective and assumed the contact method, and by their accurate and timely reports of the enemy's movements, enabled the Austrian commander-in-chief to gain advantages, both strategical and tactical, over his adversary's superior forces, and to fall on them with signal success. It is evident from these examples that the proviso of General Pelet-Narbonne—"otherwise they [divisional cavalry] would be strengthened for the purpose by the cavalry divisions"—may, in extreme cases, lead to the absorption of the whole of the cavalry on protective duties. Nevertheless, such cases are extreme, and are also rare, if only because a commander in the field can seldom obtain from other sources such complete information that he can afford to dispense, even temporarily, with that obtained by contact reconnaissance.

When the enemy has a great superiority in troops suitable for reconnaissance, the choice of the method to be employed requires careful consideration. The first temptation which assails a commander is to reinforce his independent cavalry by the protective squadrons, in the endeavour to bring the former up to such strength as will enable it to make head against its opponents. If this course offers a fair chance of immediate tactical success, it may be worth adopting, but there are grave dangers attached to it. The objects of the cavalry being, for the time, purely tactical, may draw it to some distance from the main army; in this case the army is left with no mounted troops at all for reconnaissance, and there can be no doubt that such a situation is perilous. Another possible course, of which an example (Custozza) has already been quoted, is to employ the whole of the cavalry defensively on protective reconnaissance, trusting to independent patrols and other sources of intelligence for the necessary positive information of the enemy. This procedure is sometimes effective enough, and any army which is decidedly inferior in mounted troops may, on occasions, be obliged to have recourse to it. Its effectiveness is, however, dependent on the possibility of obtaining information by other means; and even if this difficulty be overcome, there are two disadvantages to be faced. One, is that the quality of the cavalry may be impaired by a continual defensive attitude; the other, that there is no chance of reducing the superiority of the enemy by tactical success, should he give opportunities for action through carelessness or over-confidence.

Probably the best course, as a rule, is to retain a distinction between the troops allotted to the duties of contact and protective reconnaissance. If, under the conditions of the special case, more troops than the necessary minimum for either service are available, the surplus may be used to strengthen the other force. It is often possible also, to support, or even partially to replace, protective cavalry by infantry, thus setting free a proportion of mounted men. The objection to the employment of infantry in this way, in the case of a large force, is the necessity of pushing the detachments out to a considerable distance. This will render the work certainly arduous and perhaps dangerous, owing to the difficulty of retirement in case of sudden attack. Whatever plan be adopted, the security of the army must be assured by proper protective arrangements, in order to leave the independent cavalry free to manoeuvre. This latter may then be massed on either flank, ready to undertake contact reconnaissance on a limited front, or to operate tactically against any detached portion of the enemy's mounted troops.

Independent reconnaissance is used as an adjunct to both contact or protective service but is not an efficient substitute for either. For it is evident that it is only under unusually favourable circumstances that independent patrols, having got into touch with an enemy, can keep contact and remain in continuous observation. To attain these ends, combination and support are equally necessary; and it is exactly the freedom from combination and the absence of support which are essential to the independence of this type of reconnaissance. With regard to protective duties also, although independent patrols may sometimes be sufficient to guard a force from surprise, they cannot well fulfil the other requirement—that is, to protect it from the observation of the enemy. Independent reconnaissance is therefore supplementary to, and not a substitute for, reconnaissance of the other types.

The definitions and distinctions adopted here in discussing reconnaissance of the enemy are based on the methods employed, and not on the objects to be attained. For it is only in this way that reconnaissance can be considered in detail. The use which is to be made of information is of moment only to the man who wants it—the responsible commander of troops; those who are detailed to get the information are concerned chiefly with the consideration of the best method by which it can be acquired. It matters very little either to the cavalry leader or to the individual scout whether the information he is told to get should be academically defined as strategical or tactical; nor, indeed, is it always very clear to the commander-in-chief himself. These terms, in fact, although frequently used, are somewhat misleading, and even if easily understood, would not give any aid to those who wish to gain a practical knowledge of reconnaissance. For the distinction between strategical and tactical reconnaissance really depends on the result, not on the intention; a reconnaissance ordered for a strategical purpose may obtain only information of tactical value, and conversely a tactical reconnaissance may clear up the strategical situation. The introduction of such terms only complicates the subject.

On one point, however, the intention of the commander must be made quite clear: that is, whether an operation he has ordered is to be a reconnaissance or is to be something quite different. It is to be feared that commanders may sometimes find a difficulty in making up their minds on this subject, particularly with reference to the enterprises known as cavalry raids.[51] These are frequently ordered with a double object; to gain information, and at the same time to inflict damage on the enemy. But the officer who is to carry out an operation of this kind should know very definitely which of these is the more important, otherwise the raid may be conducted on the wrong principle. It is necessary for him, in fact, to know whether he is engaged on a reconnaissance, in the course of which he may find an opportunity of damaging the enemy or is engaged in a tactical operation in the course of which he may pick up some information. It is evident that his procedure must be based, to some extent, on this knowledge.

The moral and material damage inflicted on the enemy by some of the cavalry raids during the American Civil War[52] was

---

51. A *raid* is used in operational and tactical warfare to make a quick strike rather than capture a location. Some objectives of raiding are to destroy installations or stocks, free POWS, and capture key military personnel for, say, interrogation, as well as to gather intelligence.

52. Starting with the Battle of Fort Sumter on 12 April 1861 until the surrender of General Robert E. Lee at Appomattox Court House, Virginia on 9 April 1865. Though President Andrew Jackson did not officially declare an end to hostilities until 20 August 1866.

so great, considering the means employed, that the value of such expeditions in the matter of reconnaissance has become somewhat obscured. The first of these, Stuart's raid round McClellan's army in the Peninsula in 1862,[53] was initiated and conducted after the manner of a reconnaissance, and the infliction of damage on the enemy was intended to be a secondary consideration. The object of the expedition was to discover where the right flank of the Federal army was posted, and whether it offered a favourable objective for attack. Making a wide detour, Stuart rapidly closed on this flank, and found that it was practically in the air, unprotected and unsupported. His object was thus attained; but he had by this time alarmed the whole Federal army, and he suspected that measures would be taken to intercept his retreat, if he should return by the same route. He therefore decided to pursue his course round the rear of the enemy, and to inflict as much damage as he could in passing. This bold project was successfully carried out; in his three days' ride Stuart and his 1,200 men made a circuit of the whole Federal army. The raid, therefore, as a reconnaissance, was quite successful; as a tactical operation it was not only successful but original, and the moral result especially was far-reaching. As a consequence, the tactical importance of such raids was exaggerated: so much so that cavalry was sometimes withdrawn from its proper duty of reconnaissance, to proceed on expeditions on which there was no expectation or possibility of gaining useful information. This policy was to a great extent responsible for the Federal defeat at Chancellorsville. The main body of the cavalry, some 10,000 men, was sent on a tactical errand outside the sphere of immediate operations, and thus left

---

53. James Ewell Brown "Jeb" Stuart (1833–1864), was a general in the army of the Confederate States of America, and George Brinton McClellan (1826–1885), a general in the Union Army.

the superiority for reconnaissance, in the immediate theatre, with the Confederates. It was this superiority which rendered Jackson's[54] flank march possible. Such expeditions, therefore, may be dangerous not only to the troops which take part in them, but also to the army from which these troops are detached; and there is no convincing evidence that in any of these great raids, except in those which were undertaken for the special object of obtaining certain definite information, the final result was worth the risk. It is conceivable that it might be so, but it is not easy to prove a case. Possibly Stuart's raid on Pope's[55] communications in August 1862, before the battle of Manassas,[56] might be quoted as an exception. The raid was intended to effect material damage, and in this object was only partially successful. But, by a fortunate chance, valuable information was obtained by the capture of Pope's dispatch book, containing full details of the strength and dispositions of the Federal forces. This information had a direct and far-reaching effect on the issue of the campaign; while the fact that the strategical and tactical effect of the raid itself was but slight, is evident from the attitude of Pope, when, later, his communications were severed by Jackson's corps. He was then, at first, inclined to disregard the interception entirely, believing it to be only another cavalry raid, and therefore of small importance. The valuable result of Stuart's raid was a matter mainly of chance and can hardly be said to justify the risk which was incurred or to recommend the attempting of similar ventures as practical operations.

---

54. Thomas Jonathan "Stonewall" Jackson (1824–1863) was a general in the Confederate Army between 1861 and his death on 10 May 1863.

55. i.e., General John Pope (1822–1892).

56. Referred to as the Second Battle of Bull Run and the Battle of Second Manassas. It was fought between 28 and 30 August 1865.

Cavalry raids were undertaken by both Japanese and Russians during the late war in Manchuria.[57] Of these, it would appear that the Japanese raids were in no sense reconnaissances, being despatched with the sole object of disturbing the enemy by damaging his communications. Some of the enterprises of the Russians, however, seem to have included the pursuit of information in their aim, and may therefore be classed as reconnaissance operations. General Mischenko's[58] raid in January 1905, which is the only one on which authentic information has been published, was apparently designed to clear up the situation with regard to the advance of General Nogi's troops from Port Arthur to reinforce the main army, and also to disturb the Japanese communications.[59] The force employed was something under 10,000 men. General Mischenko obtained the required information and was also technically successful in damaging the enemy's communications, but the damage done was trifling, and the losses were so heavy that the expedition can hardly be considered, on the whole, a success. For there seems to be no doubt that the raid was designed, primarily, to affect the destruction of the Japanese stores at Ying-kou, and the attack on this place was a complete failure. For the accomplishment of the other objects, the force was unnecessarily large; the information about the move of 60,000 men along a single line of rail could not easily be missed by any patrol in a country whose inhabitants

---

57. During the Russo-Japanese War which was fought from 1904 and 1905 by the Russian Empire and the Empire of Japan over Manchuria and Korea.

58. Pavel Ivanovich Mishchenko (1853–1918) was a general in the Imperial Russian Army.

59. Henderson is referring to a raid in relation to the Siege of Port Arthur, which began on 1 August 1904 and lasted until 2 January 1905. The objective of the Japanese incursion was the Russian naval base and deep-water port situated at the end of the Liaodong Peninsula, Manchuria.

were neutral, and the slight damage to the railway, which was all the material result of the enterprise, might very well have been effected by a squadron or two. Had information been the vital objective of the expedition, it would have been better for General Mischenko to have enveloped, by contact reconnaissance, the left of the Japanese army, and then to have dispatched such independent parties as he thought necessary in search of the information. Such procedure would have given quite as much prospect of obtaining the information, without the risk of heavy losses to the cavalry. Which of his objectives General Mischenko considered the more important is not yet known; but if he looked on his raid as a reconnaissance, it was not well conceived; and if he thought it was a tactical operation, it was conducted with insufficient resolution.

The employment of an exaggerated force on reconnaissance, in order to deceive the enemy, is a stratagem which has frequently been used.[60] It is, indeed, almost an essential part of the program of any detachment which is employed in detaining a superior enemy by demonstration while the main force operates elsewhere. The methods both of contact and of protective reconnaissance should be employed, and a commander should not hesitate to detail his whole force, if necessary, for these duties. For the more resolutely touch is maintained, the more likely is an opposing commander to believe that the patrols are strongly supported; and the less his reconnaissance is able to pierce the protective screen, the less is he likely to find out that it has nothing to protect.

---

60. Deception in war can be traced to ancient times through treaties such as Sun Tzu's *The Art of War*. See, for example, *Sun Tzu on the Art of War*, translated by Lionel Giles (London: Luzac and Company, 1910).

An example of the strategical value of reconnaissance, when used to deceive an adversary, is found in Stonewall Jackson's campaign in the Shenandoah Valley in 1862. Ashby,[61] the leader of Jackson's small cavalry force, believed in aggressive reconnaissance, and throughout the campaign he kept touch with the main Federal force. So skillful and resolute was he that it was impossible for the Federal commander to guess, from the demeanour of Ashby's cavalry, whether they were strongly supported or not. The position of Jackson's main force was consequently a matter of doubt to the Federals at all times, save when actually in action, for Ashby followed exactly the same procedure whether Jackson's main force was one mile or fifty miles distant. On those occasions when Ashby was left to his own devices, he employed his whole force on reconnaissance, seeking unnecessary information with simulated eagerness, protecting an imaginary army with pretended earnestness, and by these means enveloping his adversaries in a fog which was lifted only to show Jackson himself at the head of his battalions secure in the possession of every advantage which skill, surprise, or stratagem could give.

The value of reconnaissance is largely influenced by the attitude and temper of the inhabitants of the country which is the theatre of hostilities. If the population be friendly or neutral, information and assistance may be obtained; if passively hostile, no assistance can be hoped for, and but little information; yet it may still be possible to carry on reconnaissance by the ordinary methods, adopting only those additional measures of precaution which are imposed by the knowledge that the enemy will have the benefit of both information and assistance. But if the population be actively hostile, the effect on reconnaissance

---

61. i.e., Turner Ashby, Jr. (1828–1862) was a Confederate cavalry commander.

becomes serious. In such case, small patrols cannot be used; all reconnoitring parties must be made up to a strength sufficient to enable them to overawe or at least cope with the inhabitants, and consequently the secrecy of movement necessary to elude the enemy becomes almost impossible. The possibilities of independent reconnaissance, therefore, are but limited, and even in contact reconnaissance the risks are much increased. The effect on reconnaissance of the alteration in the temper of a population from passive to active hostility is remarkable; and the operations of the Germans in 1870[62] give an illustration as complete as could be desired. At the outbreak of the war, the civil population of France, although of undoubted patriotism, did not actively oppose the invaders in any way. The responsibility of resistance lay on the regular army, and at that time there was no knowledge that the army would prove unequal to its task. Moreover, the inhabitants were dazed by the swift advance and repeated successes of the German armies; the crisis was unexpected, and the combination necessary for organised resistance was wanting. In a few cases, sporadic outbreaks did take place, but the strict reprisals of the invaders were sufficient to check any tendencies in this direction. For these reasons, as long as a French regular army remained in the field, the inhabitants gave but little trouble to the Germans, whose patrols and scouts, practically unmolested, were able to spread far over the country seeking out the organised forces of their opponent. Contrast these conditions with the possibilities of reconnaissance after the genius and passion of Gambetta[63] had evoked the spirit of popular resistance. Colonel Lonsdale Hale, in *The People's*

---

62. i.e., the Franco-Prussian War from 19 July 1870 to 28 January 1871.

63. Referring to French statesman Léon Gambetta (1838–1882) who was a politician during and after the Franco-Prussian War. Subsequently, he served as France's 37th Prime Minister from 1881 to 1882.

War in France,[64] gives an admirable picture. "Patrols and small parties seeking information were held back at every village; they carried their lives in their hands; the patrols would be shot down by a countryman behind a hedge; and to obtain truthful information was extremely difficult. The officer or orderly, carrying a report or an order, sometimes disappeared mysteriously; a small party of soldiers would, perhaps, be surprised at night by a few inhabitants who had noted down their sleeping quarters. When the hostile armies were so near to each other that their outposts were in contact, the inhabitants lied freely when asked about the position of their own army; and, observing closely that of the invaders, passed the information on to their own troops. The Germans, therefore, from want of the necessary knowledge, had no data on which to frame their strategical operations; these were consequently based on guesswork only; and large forces were marched off in wrong directions against a foe which was elsewhere, or which existed in imagination only."[65]

The fact that the "People's Army," in spite of these advantages, was easily overthrown by inferior German forces, does not in any way indicate that the advantages were illusions. For reconnaissance is only a means to an end; the end must be attained on the battlefield; the most perfect reconnaissance, the most inspired strategy, cannot of themselves ensure success. The

---

64. Colonel [Sir] Lonsdale [Augustus] Hale, *The People's War in France, 1870–1871* (London: H. Rees Ltd, 1904).

65. Now, more commonly referred to as *guerrilla warfare*, this is a type of irregular combat. Individuals or small groups of combatants use military tactics to engage a numerically stronger and more well equipped, but less-mobile army. Guerrilla forces are violent non-state actors who set ambushes, sabotage facilities, conduct raids, harass opposition forces with hit-and-run tactics, because they are mobile.

commander who can devise a skilful design and organise an efficient reconnaissance arrives on the battlefield with certain advantages, and that is all. He may find his enemy surprised, out-manoeuvred, and ripe for defeat. But the ordeal of battle must be undergone; and until one side has inflicted, and the other accepted defeat on the battle-field, that success which is the object of war has not been achieved. It is well for us that it is so, for it has fallen to British soldiers, more than to those of any other nation, to redeem with their blood on the stricken field the errors of blind strategy and careless reconnaissance.

CHAPTER II

PROTECTION AND SECURITY

The terms "protection" and "security," when used in a military sense, are so nearly synonymous, and the methods by which these two services are conducted are so similar, that it is not surprising that they should be looked on by most soldiers, and even by the writers of text-books, as the same. There is, however, a distinction, and an appreciation of this distinction is essential to the consideration of the details of reconnaissance, for the two duties sometimes impose a double responsibility. Protection implies two forces, one of which is protecting the other; measures of security refer to the precautions which any one force takes to guard itself from surprise or from observation. Thus, the commander of a force detached on protective reconnaissance has a double duty—to protect some other force, usually but not necessarily that from which he is detached, and also to secure the safety of his own force. In our regulations, the two duties are recognised; although the distinction is not made clear, the instructions provide generally for the safety both of the main force and of the detachment which protects it. The principle is accepted, also, that a force employed on protective duty must at all costs protect; its own security is a secondary matter, and consideration for that security cannot excuse any failure to fulfil the paramount duty. When the whole force is small, and the outposts therefore contracted, there is not usually any difficulty in reconciling the two aims; but when large forces are employed, it becomes necessary to push the protective screen far out, and in such cases the requirements of protection and of security may become sharply divergent, and even, if the force be insufficient, incompatible.

In considering the question of the strength of a force which is to be detailed for protective reconnaissance, a commander should give weight to both aspects. If he should limit the strength to the bare requirements of protection, he may render it impossible for the protective force to take any measures for its own security; the safety of the main force is then secured only at the risk of sacrifice of the detachment. The strength of a protective force should be sufficient to enable its commander to fulfil three conditions, two of protection and one of security—to give due warning to the main body of the enemy's presence or approach; to prevent the intrusion of the enemy's scouts; to obtain such warning for himself as will give him a chance of saving his own force. In the case of any except the smallest force, sufficient warning for either protection or security implies some power of delaying the enemy, a tactical consideration which must not be lost sight of.

The proportion between the strength of any protective force and the extent of front which should be allotted to it, is primarily governed by these conditions, the requirements of which vary according to the strength, proximity, and activity of the enemy, and also according to the nature of the country. A commander, when detailing such a force, must balance his desire for adequate protection against his reluctance to detach a large proportion of his force, and the suitable compromise is not always easy to fix. If, on the march, he employs large advanced and flank guards, his control is lessened thereby; if, at the halt, he should strengthen his outposts, with a view to ensuring complete safety from surprise, he will correspondingly deprive more men of their rest. Most of all will the problem be difficult when it becomes necessary to despatch a force to screen a concentration or to conceal a manoeuvre; for such duties must be entrusted mainly to mounted troops, and the whole question of the use of cavalry, offensive or defensive, inquisitive or secretive, cavalry divisions or divisional cavalry, comes at once to the front. Normally a commander will have at his disposition forces designed for both purposes, but his decision as to the sufficiency of either must be based on the actual requirements of the case with which he is confronted. The permanent allotment of cavalry in the order of battle[66] is arranged for convenience, and with reference to normal requirements; but adherence to normal arrangements in abnormal circumstances is simple pedantry, and in war is likely to lead to disaster. The strength of a force for protective reconnaissance should be estimated so as to allow for the proper performance of its duties and a reasonable chance of its safety and should be limited strictly to that standard.

Protective reconnaissance is, in the main, either passive, as in outposts, or tactical, as in advanced or flank guards, and the principles on which these duties should be conducted are so strictly included in the education of every soldier, that it is unnecessary here to dwell upon either aspect. A certain amount of active reconnaissance, chiefly for purposes of security, is, however, incumbent on forces of either type, and this subject, as well as the duty—not always passive—of a detached protective force or screen, requires consideration.

The active reconnaissance which devolves on an outpost force is practically confined to the despatch of reconnoitring patrols, and the value of these patrols depends very much on the discretion with which they are used. The vague and somewhat misleading term "reconnoitring patrol" is technically confined to

---

66. A military force's "order of battle" refers to "The method or manner in which [it] is trained, equipped, organised and disposed." Woodford Agee Heflin, editor, *The United States Air Force Dictionary* (Washington, D.C.: Air University Press, 1956), 362.

protective or security patrols, and is not applied to independent reconnaissance, which is a matter entirely separate from the question that is now being discussed. A reconnoitring patrol is one which is sent out to ascertain the absence of the enemy from a particular area; on the completion of its limited exploration it returns to its lines. Very often, no doubt, the ground in front of an outpost line is of such a nature that periodical investigation in certain directions is necessary, and in such cases reconnoitring patrols must be used; but here are serious objections to the indiscriminate application of this system of reconnaissance in front of an outpost line. These objections particularly apply to the patrols sent out from infantry outposts. Their range is so limited that the additional warning they give can be of but little value; the fact that patrols are in front lessens the sense of responsibility of the sentries in the fixed observing line, and also causes doubt as to whether any person or party approaching the line is an enemy or a patrol returning. Any doubt of this kind leads to the risk of patrols being shot down by their own troops, especially at night. Even if fire be withheld, the doubt is likely to be communicated to the troops and cause false alarm or unnecessary preparation. On the night before the action at Omdurman in 1898,[67] a native patrol returned somewhat hurriedly to the zareba[68] of one of the British brigades, with the result that the whole brigade stood to arms and lost an hour's rest before an exhausting day. If patrolling to the front were adopted as the regular procedure of the outposts of a large force, such incidents would be continually occurring. The system is impressive enough on peace manoeuvres, but in war, if the

---

67. The Battle of Omdurman took place on 2 September 1898 between the armies of General Sir Herbert Kitchener (Britain) and Abdullah al-Taashi (Sudan), who was defeated.

68. This is a thorn fence structure that is used to fortify a camp.

enemy be within striking distance, patrolling of this kind should be very strictly limited.

Mounted patrols in advance of an outpost line are more practical than those composed of infantry. Their range is wider, their reports are more quickly transmitted, their retreat can be more rapid, and they are less likely to be mistaken for an advancing enemy. Night work, however, is very exhausting for horses; at night it is usually better to lay the whole responsibility of observation on the fixed outpost line. By day, the judicious use of cavalry patrols may lighten very appreciably the strain of outpost duty. If the country be suitable, cyclist patrols[69] may also render good service by day; and by night, if patrolling be absolutely necessary, they would appear to possess some advantages over patrols of either cavalry or infantry, owing to the silence and swiftness of movement which the cycle confers.

Reconnoitring patrols on the flank of an outpost line are evidently not open to all the objections which apply to those despatched to the front. In many cases they may be of great value, especially when the force for outpost duty is limited.

The reconnaissance which is carried out by advanced guards is only partly protective in its nature. For an advanced guard must be prepared to fight, either offensively or defensively, at any moment. Its reconnaissance, therefore, besides providing for protection and security, should be conducted with a view to obtaining tactical information. When an advanced guard meets the enemy, information about that enemy is of vital importance to the advanced guard commander.

---

69. Over the last century, the types of reconnaissance vehicles expanded to include many categories of specialised vehicles—motorcycles, four-wheeled and six-wheeled lightly armored vehicles, light tanks, jeeps, and different configurations of fighting vehicles.

If he is not informed, he may shatter his force against the enemy's main army, or he may be daunted and delayed by the bold attitude of a few patrols. The study of advanced guard actions in past wars leads inevitably to the conclusion that the value of these forces depends much more on the efficiency of their reconnaissance than on their fighting strength. Information sufficient to enable an advanced guard commander to decide correctly when to attack and when to defend is worth many battalions.[70]

The accepted methods of advanced guard reconnaissance are mainly protective, and to this purpose they are well enough adapted. Patrols are sent to the front for the security of the advanced guard, and to the flanks for the partial protection of the main body, and if well handled, these should usually be able to give the necessary warning. But the wedge-shaped formation which a vanguard under this system usually assumes, however convenient for protective purposes, is not calculated to produce the best results in the matter of information. For it is entirely a defensive formation, and defensive principles are not universally applicable even to protective reconnaissance. In the case of an advanced guard, the best protection is given by the detection and examination of the enemy, and for this purpose it is advisable to keep the observing parties on the flanks well forward, strengthening them so that they may be able to take measures for their own security in their possibly exposed situation. In fact, for the purpose of acquiring information, the mounted troops of the advanced guard should move on as broad a front as their numbers and the nature of the country will allow, approximating their methods to those of a force engaged in contact reconnaissance. Especially should this be the case when several

70. A succinct summary as well as an important axiom of reconnaissance.

columns are moving on parallel roads: there is then a possibility of common action and mutual support. In any case, the protection given to the main body by patrols thrown back on the flanks of the advanced guard can be only partial, and in most cases this purpose would be as well served by stronger bodies which maintained a more forward position.

To make the general contention clear, let it be imagined that two opposing forces, each with a vanguard of mounted troops, are approaching each other on the same road. One of these vanguards is in the ordinary wedge-shaped formation, with the main strength in the centre, and patrols thrown back in the flanks. The other is in three bodies of nearly equal strength, in line, or with the flanks slightly forward. When contact is made, that force which is strong on the flanks is certainly, from a reconnaissance point of view, the more favourably situated, for, even although it is tactically inferior in the centre, that is exactly the place where the enemy will acquire least information, as his advance will be checked by the main guard. The strong flanking parties, on the other hand, if pushed forward, as they ought to be, should be able to overcome the hostile patrols, and obtain some definite information as to the strength and composition of the enemy's main force.

The insufficiency of the wedge formation for advanced guards is shown by the fact that, when, owing to the nature of the country, it becomes difficult to ensure protection, or security, the formation is almost invariably abandoned. An advanced guard approaching a defile such as a mountain pass, is halted, and the flank patrols pushed forward to examine the heights on either side. Similarly, in approaching woods or villages which lie off the track of the column, it is customary to send patrols forward to see that these dangerous spots are clear. An arrangement which must be modified in this way to achieve the very object for which it was designed is not quite practical; if the flanks must be advanced to cope with difficulties, it would seem better to keep them advanced throughout.

When the mounted troops of the advanced guard are insufficient to cover a front as broad as the commander thinks necessary, it may even be advisable to keep them entirely on the flanks, where they can best obtain information, leaving the security of the immediate front of the column to the infantry. The method to be employed by the cavalry of an advanced guard approximates to that of contact reconnaissance, with the advantage of close infantry support. This support renders it possible to make reconnaissance the principal aim; the assistance of advancing infantry is a factor which modifies the necessity of providing for tactical cavalry action on a large scale. The cavalry may therefore be dispersed to a greater extent than would be advisable in an ordinary contact reconnaissance, and this dispersion makes it possible to cover a broad front, and, at the same time, to examine thoroughly the ground traversed. If this deployed mounted force be checked by the enemy, it should not concentrate for cavalry action, but should, if possible, continue its active reconnaissance, calling on the infantry for support. To enable the infantry of the advanced guard to render assistance speedily, if is desirable that they too should move on a broad front—that is, that there should be detachments on the flanks in immediate support of the cavalry. If the whole of an advanced guard moves on one road, save only for patrols on the flanks, it is evident that deployment will be necessary to drive in any opposition which may be met with. It would therefore seem reasonable to retain on the march, when possible, a formation which will aid rapid deployment. The extent to which this can be done depends chiefly on the nature of the country, but in most cases, it will be possible for close support to move on the flanks at such interval from the main guard as will ensure the speedy reinforcement of any part of the vanguard. There may be some tactical objection to such a formation, but, if so, it is not easy to discern; there may even be, in a narrow and pointed front, some hidden virtue which no encounter has proved, and no writer has claimed; but considered with reference to the acquisition of information, and to the avoidance of delay, a broad and well-supported front is, for a vanguard, a formation which has many advantages.

The protective reconnaissance carried out by flank guards and rear guards does not call for consideration in great detail. Regarding the security of a moving flank guard, it may be suggested that the best method of ensuring it is by the successive despatch of security patrols to the flank. The system frequently adopted of detaching patrols or stronger bodies to move parallel to the flank guard, but still farther to the flank, is not suitable for reconnaissance purposes. The restrictions imposed by the necessity of moving at a certain pace, and of observing always towards a flank, limit seriously the possibility of efficient scouting, as will be noticed later when the details of scouting are considered. It is better that patrols should move towards the probable enemy—that is, at right angles to the direction in which the flank guard is moving; that they should be despatched at such intervals as will permit of the exposed flank of the guard being continuously covered by observation; and that each patrol should, when the flank guard has passed, re-join it by its rear. This method of successive patrols employs rather more men and horses than are required for the single parallel patrol, but the increased security which it confers entirely outweighs this consideration.

The reconnaissance which should be attempted by the rear guard of a retreating force cannot very well be estimated in a general way, for the possibilities must depend very much on the conditions of the case. There is one precaution, however, which is almost invariably applicable—that is, reconnaissance extended widely to the flanks. For a pursuing enemy, if he knows his business, will certainly endeavour to turn a flank of the rear guard, either with a view to intercepting the rear guard itself, or with the intention of disregarding the rear guard, and harassing, even attacking, the main body. In the first case, reconnaissance to the flanks is necessary for the security of the rear guard; in the second, it is a service of protection for the main body; in both, it is a precaution eminently necessary for the safety of the army.

A protective screen is employed either as an adjunct to the ordinary outposts of a force, or, more commonly, to cover and conceal a strategical or tactical manoeuvre. Such a screen is usually composed of mounted troops, but sometimes infantry is used to strengthen it; and it is conceivable that, in favourable country, the duty might be entrusted entirely to infantry. A protective screen is designed not only to give warning of the enemy's approach, but also to deny information to the enemy, and it must be the aim of the commander to block, absolutely, the passage of the enemy's scouts. In many cases, it will be necessary to stop communication altogether, even the inhabitants being prevented from crossing the line in either direction. The measures to be taken to ensure this end depend a great deal on the nature of the country, and there are advantages in deploying a screen along a natural obstacle, such as a river. The usual and most economical method of stopping communication, is by a system of posts, between which patrols move at irregular intervals; but the conditions under which duties of this kind are carried on vary so much, according to the force available, the orders which the commander has received, and the ground which must be watched, that few principles can be generally applicable. One of the most important measures, both for protection and for security, is the organisation of rapid communication all along the line. Good telegraphic communication is invaluable;[71] it will save horses and men and will make it possible for the commander to hold his line lightly, and perhaps to retain reserves for tactical employment at any threatened spot. Patrolling, both protective and independent, to the front, in addition to the linking patrols between posts, is useful as a means of gaining extra warning of any suspicious movement of the enemy. Most important of all, is the preparation of simple and unmistakable instructions as to the manner in which the duty is to be carried out, so that every officer and man shall understand and appreciate his duties and responsibilities. This is all the more necessary if the protective screen is intended to act as a barrier against the secret, as well as the open endeavours of the enemy to gain information.[72] In such case, the movements of the inhabitants must be limited, or even stopped entirely; and whenever military restrictions are imposed on inhabitants, it is desirable that they should be applied with uniformity, and not at the discretion of subordinates. For there is no subject on which the principles or the predilections of soldiers vary so much as on that of the proper treatment of civilians in war.[73] Strict

---

71. Battlefield communications are as old as warfare. Methods such as flags, drums, runners and messengers on horseback were some of the early forms used. However, as the decades of the twentieth-century advanced, radiotelegraphy and radiotelephony came to dominate military field communications. Now, encrypted digital communications have replaced analogue forms of wireless communication. See Peter R. Jensen, *Wireless at War: Developments in Military and Clandestine Radio, 1895–2012* (Dural, New South Wales: Rosenberg Publishing, 2013).

72. The protection of secrets falls into the branch of intelligence known as *counterintelligence*. See a detailed discussion of all forms of information security in Hank Prunckun, *Counterintelligence Theory and Practice, Second Edition* (Lanham, MD: Rowman & Littlefield, 2019).

73. An important piece of advice that is as important as learning any tactic. For those who might doubt the importance of this thinking, I refer
–continued–

regulations and strict supervision are necessary in order to ensure anything like uniformity of treatment, and it is only by the impartial and universal application of restrictions that discontent, and friction can be avoided. In some wars, regulations imposing restrictions on the inhabitants have been used by subordinates as instruments of personal oppression and violence;[74] but the British army has never erred on the side of laxity in their application, forgetting that such regulations are made in the interest, not of justice, but of expediency. Occasional cases of individual hardship, caused by the rigid enforcement of regulations, are by no means without effect in impressing on the population the military importance attached to the rules, and in thus inculcating the belief that they are unavoidable and had better be accepted. But any difference of opinion among the military authorities, any local relaxation of the rules, will at once dissipate this belief and cause active discontent.

---

to the numerous examples found in the news media of soldiers who have been prosecuted for crimes committed during war.

74. Take for example, the methods used by occupying Nazi forces during the Second World War and the members of the Islamic State in Syria during the mid-2010s.

CHAPTER III
CONTACT RECONNAISSANCE

Contact reconnaissance may be defined as the endeavour to secure information by pushing troops close up to the enemy on a broad front, pressing back and possibly breaking his advanced screen, and holding his main force under observation. It is the chief duty of the independent cavalry of an army, and the greater part of the training of modern cavalry is designed to fit officers and men for the work. In the extreme application of contact reconnaissance, the enterprises called reconnaissance in force, the three arms may be usefully employed; and in cases where the opposing forces are in close contact all along the line, infantry may well take part; but as a rule, the duty falls to the cavalry, and the subject must therefore be considered principally with reference to cavalry conditions.

It is first necessary, however, to clear the way by investigating the rather obscure subject of reconnaissance in force, a term which has frequently been misapplied.[75]   A reconnaissance in force properly means the employment of a considerable portion of the whole available force for the purpose of gaining information about the enemy or about the enemy's

---

75. The following are examples of how terminology has been misused by those ignorant of the word's meaning, thereby corrupting the term. Two illustrations present themselves: the first is the term *tactical intelligence*, which has incorporated the notion of operational intelligence. Take the case where an analyst produces a "tactical assessment," when in fact the report should be termed an *operational assessment*. The other example is the now widespread misuse of the term "social engineering." The term refers to societal scale human manipulation, but in computing environments it has been corrupted to mean a ruse or pretext.

position; and the use of the term implies that the commander must engage, or at least offer to engage, the enemy for that purpose. Minor operations of a similar type are frequently initiated and carried out as ordinary details by the troops engaged in contact reconnaissance; but a reconnaissance in force connotes the employment of forces additional to those already detailed for permanent reconnaissance duty. Frequently, however, in British warfare, the term reconnaissance in force has been applied to operations differing widely in nature from the definition here given; at one time it has indicated a demonstration to attract the enemy's attention; at another it has been used as a euphemism for an abortive or an unsuccessful attack. In some cases, also, reconnoitring expeditions, for which the employment of a large force has been necessary for defensive reasons—that is, for the protection of the expedition itself against possible attack—have been so designated; although, in its technical sense, the action of a reconnaissance in force is essentially offensive.

The occasions on which a commander in the field is likely to find it necessary to have recourse to reconnaissance in force are not many; nor is it probable that conditions favourable to such an enterprise will often be found. For the operation is in its nature tentative and indecisive; if timidly conducted, it can hardly attain its object; if resolutely pushed, the troops may become so involved that extrication is difficult. If, however, the troops to which general reconnaissance has been entrusted prove unequal to their task, owing to their numerical or moral inferiority, or possibly owing to mere difficulties of ground, it may be considered necessary to launch a force of a strength sufficient to induce the enemy to expose his main dispositions. But the operation is a delicate one. Demonstrations at long range will not usually tempt a disciplined enemy to display his force; and if it be of vital importance to discover his arrangements, there may be no choice but to deliver an attack with an inferior force, in ignorance of the nature or extent of the opposition to be met with. Such an operation is undoubtedly hazardous, and even if the required information be obtained, the losses may be heavy, and the moral effect of the encounter, which will be looked upon by the troops on both sides as an unsuccessful attack, may be considerable.

In the conduct of a reconnaissance in force there are certain elements which are almost essential to success. Surprise may do much to disconcert an enemy and to tempt him to action on an unnecessary and enlightening scale. To gain the full effect from surprise, swift and resolute action is necessary, and there must be no hesitation either in the advance, or in withdrawal when retreat is decided on. Most important of all, perhaps, is the choice of ground, when choice is possible. Not infrequently the features of the ground in the vicinity of the enemy may be such that a reconnaissance in force may be made a part of a meditated attack; that a real, but limited attack may be delivered on an advanced position of the enemy, so that the reconnaissance may, if circumstances prove favourable, be converted into a preliminary advance to effect a lodgement in or near the enemy's lines. There is, in such case, the possibility that retirement, which is the main disadvantage of reconnaissance in force, may be unnecessary; but this can hardly be the case unless the commander-in-chief is prepared to engage his whole force. And, in fact, the study of operations of this nature leads to the conclusion that a commander who fears a general action should not indulge in such enterprises; if isolated and unsupported, they can result only in losses and moral depression.

When the troops engaged on reconnaissance work are unable to pierce the enemy's screen, the manner in which they should be supported depends on the intentions of the commander-in-chief. If he is prepared to accept battle, he may employ a reconnaissance in force; if not, he must confine himself to supporting his advanced troops by such bodies of the other arms as may be required to give them the necessary confidence and weight to cope with their opponents.

When the independent, or contact cavalry is launched on reconnaissance, its main object should be the acquisition of information regarding the enemy's main forces. But in nearly every case, the effective pursuit of this aim cannot be even initiated until a preliminary object has been attained—that is, the overthrow of the hostile mounted troops. It is possible that the enemy's cavalry may be too weak, or too timidly handled, to offer serious opposition; or it may be employed on another flank, or at a distance so great that it cannot for a time be brought to action; but, as a rule, its opposition must be reckoned with from the first, and its discomfiture must be considered a necessary prelude to success in the principal task.

The contact cavalry must therefore undertake two successive reconnaissances—one to gain information of the enemy's mounted troops, and, when these have been disposed of, another to gain information of his main forces. The first is a reconnaissance for tactical, the second usually for strategical purposes; and the operations of which they form part are frequently classed as the tactical and the strategical employment of cavalry.[76] These terms are perfectly accurate, and for the purposes of explaining the principles of cavalry operations they

---

76. Here, as well as elsewhere in the book, Henderson uses the term *tactical* to mean *operational*, and the term *strategic* to mean *tactical*. Operational refers to immediate encounters with an opposition force. Tactical refers to short-range, or time-limited engagements that are longer in duration than operational. Whereas strategic refers to long-time, future engagements that involve political and well as military policy/planning. See Hank Prunckun, *Methods of Inquiry of Intelligence Analysis, Third Edition* (Lanham, MD: Rowman & Littlefield, 2019), 15–18.

are necessary; but for the study of reconnaissance, the distinction is not required, for the method of reconnaissance employed is in both cases of the same type—contact reconnaissance.

The operations against the enemy's mounted troops may be divided into three phases: the leader of the independent cavalry must first gain information of the enemy; he must then manoeuvre to take his opponent at disadvantage; and, finally, he must attack and defeat him. A certain amount of information about the strength of the hostile cavalry, and about the locality in which it may be found, will probably be at his disposal, from intelligence sources, from the beginning; but for tactical information which is to be of use to him, he must depend on his own resources; and to the acquiring of this information his first efforts in reconnaissance must be directed. It is evident that he cannot expect to gain information sufficient for his purposes from independent patrols. These may be useful in finding out where the enemy is and where he is not; but there is little hope of their penetrating the secrets of a large mounted force so deeply, or of conveying their information so rapidly, as to justify a decisive tactical operation. Nor is the protective reconnaissance of advanced guards sufficient to ensure superiority in information over the enemy. The only reasonable hope of securing the advantage given by superior information lies in the employment of contact methods—a broad front, strong supports, concerted action. When the enemy is met with, he must be pressed continuously, until the firm resistance of his formed bodies is felt. He must be observed at every point, pressed back where he is weak, held where he is strong, while the commander considers his information and manoeuvres his masses for the combat. If his reconnaissance on these methods has been well conducted, he may hope to come into action with a tactical advantage.

It may be objected that the principles here advocated are in opposition to the accepted theory that cavalry in the preliminary stages of reconnaissance should move in concentrated formation. This objection really turns on the meaning which is applied to the word concentrated. The reasonable interpretation is that, in this case, it signifies such a disposition of the cavalry forces as will enable them to undertake combined tactical action in the most advantageous manner and at any moment. There is nothing in this incompatible with movement on a broad front, and even if there were, there is another aspect of the case to be considered. Detachments for reconnaissance purposes are always necessary, and there is ground for the assertion that, with moving forces, a strict concentration and a narrow front necessitate, for purposes of information and security, larger detachments than are required by forces of the same strength in wider formation and on a broader front. If a commander likes to move his mass of cavalry against a mobile enemy by a single road, he complies with the letter of the rule directing concentration, but he will neither obtain information nor secure safety from surprise unless he makes very large detachments for reconnaissance. If, however, he should find himself able to move his force on two or more roads or parallel routes, then the support of the columns will so strengthen the reconnaissance line that smaller detachments will be sufficient. Colonel Cherfils, in his exhaustive treatise on the employment of cavalry, is a firm supporter of the theory which favours movement on a single route. He has worked the whole question out by mathematical calculation, and his utmost concession to breadth of front is an admission that it is permissible, although not advantageous, to march a cavalry division of three brigades, one of which is advanced guard,[77] on

---

77. An advance guard can refer to several configurations of soldiers who precede the main army. Advance guard units can be tasked with identifying gaps in the enemy's defenses; eliminating minor opposition that might present as harassment/slowing tactic; and preparing the
–continued–

two roads, provided these roads are never more than two kilometres[78] apart. This, be it said, refers to a route march, not to a tactical advance. For a corps of four divisions, he allows three roads, and in his calculation assumes that these parallel routes are four kilometres apart; that is, that four divisions cover a front of five miles. The calculation is based on the distance to be traversed by the tail of the farthest column, in case of a central concentration, and on this alone; the factors of reconnaissance, of manoeuvre, of the delaying power of fire action, are not included; the consideration even of flexibility is ignored. The argument is, in fact, a little pedantic, and smacks of the study rather than of the saddle; indeed, when the author proceeds to place an imaginary force on a real map, he at once oversteps the limits which he himself has laid down.

There can be no advantage in attempting to fix any rigid rule by which to bind a cavalry commander to particular intervals or distances in the disposition of his force. Effective reconnaissance requires breadth of front; decisive tactical action demands concentration, and probably depth of formation. The commander must reconcile these conflicting influences according to the circumstances of the moment; he must decide their relative importance at each stage of his operations. Success in the leading of cavalry cannot be ensured by the application of formulae; of no art can it be said with more certainty that circumstances alter cases.

When the enemy's cavalry has been disposed of, the original method of contact reconnaissance may be employed to

---

battleground for the operation-in-chief, especially by conducting reconnaissance.

78. One kilometer is approximately 0.6 miles, so 1 mile is equal to 1.6 kilometers.

locate his main forces. The extent to which the formation may be modified by further deployment, in order to cover a wider front, depends chiefly on the completeness of the cavalry success. Formed bodies must still be kept concentrated to deal with any renewed efforts of the defeated troops, and to operate against the protective screen of the enemy. The distribution must depend absolutely on the conditions of the case.

The formation in which a force engaged on contact reconnaissance advances towards the enemy is usually that of a line of patrols or scouts, supported at intervals by formed bodies of moderate size (contact squadrons), while farther to the rear, at such intervals and distances as circumstances may dictate, follow the main bodies of the force. The patrols are intended to find the enemy, the contact squadrons to oppose him at any points where the patrols are held up or driven back, and the main bodies for tactical operations on a large scale against the enemy's reconnaissance or outpost troops. The object of every commander of a contact reconnaissance force should be to press back, or drive in, or break through the enemy's advanced troops, and obtain information at first-hand about the formed bodies of the main hostile force and having once gained "touch" in this way, he should endeavour to keep it. For keeping touch, after the enemy's mounted troops have been disposed of, the formation indicated above is suitable also, and is therefore a sound normal formation for a force thus engaged, being, as a rule, interrupted only by the concentrations or manoeuvres necessary for tactical action.

The line of supports is the backbone of this formation, and on the original deployment of this line the convenience of the formation depends. Each support should have allotted to it such breadth of front as it can effectively cover, allowance being made for inevitable temporary expansion or contraction, and, as much as possible, for difficulties of ground. The effective covering of the front means that the support and its contact patrols should keep touch with the supports on its right and left and should make certain that no formed body of the enemy is left behind on the ground passed over. The intervals between supports depend therefore not only on the nature of the country, but also on the intensity of the opposition to be expected; for if a support be given too large a front to cover, the detachment of the necessary patrols may leave it too weak to support them effectively. Such conditions cause delay, if nothing worse; and there are so many unavoidable circumstances which cause delay, that failure to anticipate a preventable cause is inexcusable.

Delay is always caused, for instance, by the necessity for communication all along the line. It is only under the most favourable circumstances, as in perfectly open country, that contact patrols on the move can maintain this communication. As a rule, the duty must be carried out by the supports, and the only satisfactory means, if the country be at all intricate, is by a system of periodic halts, on certain prearranged lines, along which communication must be established before a fresh move is made. In other words, the advance should be by stages, and each stage should be made good before entering on the next. This system in its entirety is, of course, only applicable before contact with the enemy is actually made; it must be dropped at once if serious tactical considerations intervene. For the advance towards the enemy, however, it is almost essential, if a commander hopes to get the best value from this method of reconnaissance. The essence of the method is a simultaneous advance on a broad front; every endeavour should be made to keep the line elastic; but unless occasionally reformed, it is bound to break at some point.

Lines which include some strategical or tactical points of importance, if such exist, should be chosen for the periodical halts. For it is desirable that such points should be reached without delay, lest the enemy should anticipate the movement. Strategical points also are usually, and tactical points frequently, centres of good communication, and are therefore entirely convenient for this purpose.

Contact patrols must, of course, keep communication with their supports, for at any moment they may have to send information to, or fall back on, or call for assistance from them. Besides reporting positive information, patrols should be instructed to send negative information—that is, to report the absence of the enemy, at regular intervals, either of time, or of distance traversed. During an advance, these reports should be frequent, although they are seldom urgent; at the halt, they may be at longer intervals, for information is passed more quickly and certainly when a force is halted than when it is advancing, owing to the more accurate knowledge possessed by each body of the exact position of the others.

The difficulty of keeping touch with the enemy is almost as great as that of originally gaining touch. Sometimes, indeed, it is more so, for the mere knowledge that a reconnaissance is in touch with his force may draw the attention of a commander to the slackness of his own reconnaissance which allowed his adversary to gain such an advantage, and he will probably take measures to drive off his unwelcome visitors, and to keep them at a distance. In many cases, also, a reconnaissance seeking an evasive enemy may make touch by surprise, but may find it very difficult to keep it, owing to a further move of the enemy, and to the precautions taken by him to conceal the move. Additional difficulties are presented by the task of maintaining or relieving men and horses in close contact with the enemy; positions for advanced posts must be carefully chosen and supports judiciously distributed. Nearly all the advantages of contact reconnaissance are lost if the force which has made touch be altogether withdrawn at night. Patrols must usually be retired, to some extent, to positions which offer some security; but it may well be considered a general rule that supports should not give up ground unless driven back. Limited changes of position, either for tactical reasons, or for convenience in billeting or bivouacking, are frequently necessary; but any idea of falling back on the main body, until forced to do so by the action of the enemy, is foreign to the principles on which contact reconnaissance should be conducted. The pressure on the enemy should be continuous, and even by night patrols must be constantly pushed up to his positions to make certain of his quiescence, or to give warning of his movement.

When contact reconnaissance attains complete success, the enemy is "picketed": that is, he is so enveloped by a chain of watchful scouts that every movement is under observation. This supreme development can only be attained when the superiority, especially the moral superiority, of the reconnoitring cavalry is undisputed. It was practically attained by the German cavalry on certain critical days in 1870; it was undoubtedly attained by the Boers during certain operations of the South African War. Short of actual defeat, there are few experiences more unpleasant to a commander than the sensation of being picketed. Not only are his dispositions plain to the enemy, but his own efforts at reconnaissance are usually frustrated, unless undertaken on a large scale and at some risk. For the ideal picketing line is an elastic cordon of vigilant scouts, who will retire whenever pressure is applied, and will close in whenever pressure is withdrawn, informing their supports of every move of the enemy, so that detached parties may be fallen on, and either cut off or herded back to their main body. Picketing is an art in which savage or semi-civilised warriors excel.[79] The Spaniards were

---

79. The reference to "savage or semi-civilized" people was used to describe racial differences of indigenous people. These classifications
–continued–

picketed by the Moors on more than one occasion in the wars of the fifteenth century, notably, to their great discomfiture, in the Axarquia[80]; and the Great Captain, de Cordova,[81] learning the value of this method of warfare from one enemy, applied it to another, and persistently picketed the French armies in his Italian campaigns.

Nor did the Spaniards forget the art. During the War of the Spanish Succession,[82] General Stanhope,[83] with some 5,000 British troops, left Chinchon, between Madrid and Toledo, on December 3rd, 1710, intending to retire on Aragon. On December 5th, he was overtaken by Don Joseph Vallejo,[84] a skilful partisan, with 1,200 irregular horse; and as Stanhope undertook no active reconnaissance, Vallejo was able to encompass the British force, and to picket it effectually for three days. So closely was Stanhope's force kept under observation, and so complete was the cordon, that no information of the

---

were not grounded in logic or based on scientific precepts and appear to have their origin in a paper delivered by Reverend Frederick W. Farrar who couched his theory in pseudo-science. His paper titled, "Aptitude of Races," appeared in a London journal in 1866. His theory that there are three racial categories–savages, semi-civilized, and civilized—is nothing more than racism delivered by ignorance. Today, we acknowledge such categorisations for what they are; baseless, offensive and demeaning. See Frederick W. Farrar, "Aptitude of Races," *Transactions of the Ethnological Society of London*, Volume 5, 1866, pp. 115–126.

80. Axarquia is in the administrative region of Andalusia in southern Spain.

81. Gonzalo Fernández de Córdoba (1453–1515).

82. The War of the Spanish Succession was fought between 1701 and 1714. The issue at the center of the conflict was who had the right to be king of Spain.

83. James Stanhope, 1st Earl Stanhope (abt. 1673–1721).

84. José Vallejo y Galeazo (1821–1882).

march of the main Spanish army reached the British commander, and his whole force was intercepted and forced to surrender at Brihuega, on December 9th. Lord Mahon, in his *History of the War of the Succession in Spain*,[85] offers a quaint justification of his ancestor's failure. "He had no suspicion whatever that any enemy besides Vallejo's partisan horsemen was within several marches; and relying for information on the great rewards he had promised the peasants who should bring any, he did not place on the neighbouring heights any outposts or advanced guards which might have given notice of Vendôme's[86] approach. Nor does this omission appear to me at all negligent or blameable in Stanhope, since, with a body of 1,200 partisan cavalry hovering around him, he could scarcely in any case have stationed such outposts with advantage or with safety. If stationed near the town, they would see little or nothing to report; if at a distance, they would be attacked and cut off by Vallejo." General Stanhope advanced the same plea in his report. "On the 8th, at about eleven o'clock, there appeared some horse upon the hills near the town; upon which I ordered out a party to reconnoitre; but the enemies thickening, we thought it to no purpose to send it, because we knew they might have the same 1,200 horse which had observed us, and we could not send out half their number." This sentence alone is sufficient evidence of the moral effect of successful and persistent contact reconnaissance. It was the moral, not the numerical inferiority of Stanhope's cavalry that paralysed its action. If the 1,200 men were concentrated, they might have

---

85. Lord Mahon (Philip Henry Stanhope), *History of the War of the Succession in Spain* (London: John Murray, 1836).

86. Duke of Vendôme, Louis Joseph de Bourbon (1654–1712), was a Marshal of France. History records him as one of the most successful French military commanders during the War of the Grand Alliance (also known as Nine Years' War, 1688–1697) and War of the Spanish Succession (1701–1714).

been evaded; if they were dispersed, they might have been attacked. Swift and resolute action might have broken the net; but when troops have been held continuously under observation for even a short time, their ignorance of the enemy's dispositions and the knowledge that their own movements cannot escape notice engender feelings of irresolution and even of hopelessness. In this case, it is plain that the General himself shared these feelings; and his tame acceptance of inferiority in reconnaissance foreshadowed his discomfiture in battle.

In modern times, troops trained on the European system operating in uncivilised countries are frequently picketed, as were Hicks Pasha's[87] force in the Soudan[88] in 1883 and our troops in the Tirah[89] in 1897[90]; but the drawback to the irregular method of picketing is that in many cases it partakes of the nature of an investment, and the whole army is frittered away in establishing the cordon, thus reducing the operations to petty warfare. The real value of picketing, when the opposing forces approach equality in strength, is only obtained if the duty can be performed by a suitable proportion of the whole force, leaving the main body available to utilise its advantages by decisive action. And the advantages are great; they represent the extreme possibilities of successful reconnaissance.

Picketing may be quite effective although the cordon is incomplete. What is necessary is that the enemy's force should

---

87. Colonel William Hicks, who was also known as Hicks Pasha (1830–1883).

88. This is the French name, and former English name, for the country of Sudan.

89. Tirah is situated between the Khyber Pass and the Khanki Valley in Pakistan.

90. Referring to the Tirah Valley, the location of the British "Tirah Campaign" in 1897.

be continuously under such close observation that no serious movement can escape detection. This may be accomplished without enveloping the whole force of the enemy. Contact well maintained on his flanks may be sufficient; and it may be noted that the flanks are usually the parts most vulnerable to contact reconnaissance. Contact on the enemy's rear can only be maintained under exceptional circumstances, for it means the severing of his communications—a discomfort to which no enemy will lightly submit. Reconnaissance in this direction is likely, therefore, to be confined to enterprises of an independent and temporary nature.

A partial application of the methods of contact reconnaissance is sometimes convenient for ascertaining whether a certain tract of country is or is not occupied by the enemy. The system consists in employing a line of contact patrols only, without any supports. It is suitable for reconnaissance in those directions in which the presence of the enemy, although possible, is unlikely. If any opposition be met with, the system at once becomes impossible; either the line must be supported, or its constituent parts must be broken up and the reconnaissance continued on the independent system. The value of the method lies in its economy; a weak line of this kind is just as well able to acquire negative information as is a force properly organised to pursue contact methods; thus, the absence of the enemy from large areas may be quickly and definitely ascertained by a small proportion of the total force available for reconnaissance. The system is, however, a difficult one to carry out practically; to ensure smooth working, the exact details of procedure must be laid down beforehand, and those who are to take part in it must have a clear knowledge of their duties.

The first essential is that there shall be frequent communication all along the line, and for this purpose the system of proceeding by stages is even more necessary than in contact

With Analytic Annotations

reconnaissance. The stages should therefore be short, and at each general halt every patrol should communicate with its neighbour on right or left. The report, "All clear as far as . . . ," should be passed from flanks to centre for the information of the commander of the reconnaissance; and the messengers who pass this report from patrol to patrol should carry back the commander's order, "Move on at . . . o'clock," the order being issued when all reports have been received, and the time being calculated to allow of the order reaching the flank patrols.

The necessity for frequent communication, and the advantage of moving quickly, in this kind of reconnaissance, renders it a suitable employment for cyclists, and even if sufficient cyclists to undertake the whole task are not available, the assistance of a few, for communication purposes, is very valuable.[91] In an enclosed country, with good roads, the cyclist has a distinct advantage over the horseman in reconnaissance, and for work of this nature the cyclist is peculiarly suitable, owing to his mobility. In any civilised country,[92] indeed, the cyclist is always a useful adjunct to the horseman, and for many reconnaissance enterprises may well replace him.

The conclusion which is to be drawn from consideration of the possibilities of contact reconnaissance is that, to attain success, superior force is essential. Yet there is nothing in military science so difficult to assess as the value of any force of

---

91. Contrast Henderson's urging of "frequent communication" with field communications today, with reliable voice and data wireless communications that are virtually impossible to decrypt. See Marko Suojanen, *Military Communications in the Future Battlefield* (Boston: Artech House, 2018).

92. Referring to (various forms of) "roads" in Europe that existed at the time but noting that the standard in 1914 was likely to be much lower than what would be considered a road today.

cavalry. Numbers alone do not constitute superiority; physical efficiency, training, morale, are all factors which must influence an estimate.[93] Most of all does the value depend on the quality of the leader; disparity of squadrons matters little when on one or other side there is the personality of Seydlitz,[94] Murat, or Stuart. Even if the enemy should have such a preponderance of mounted troops that his superiority must be admitted and accepted, there is yet a possibility of securing good results from contact reconnaissance. It is impossible for the enemy to ensure superiority everywhere; in modern war the flanks of an army may be a hundred miles apart, and an inferior force well-handled should be able to assert a local superiority at some point. The possibility of such success is dependent principally on the possession of one advantage—superior information—and this can only be assured by the early establishment of an efficient system of reconnaissance. That is, the better information a commander has, the more likely is he to get more.

---

93. This advice should be noted because it is human nature to "want more," and on the battlefield it is unlikely to be any different. But, as history shows, efficiency, training and morale have seen inferior forces repeal and, in some cases, defeat a numerically larger opposition.

94. Friedrich Wilhelm Freiherr von Seydlitz (1721–1773) is acknowledged to be one of Prussia's greatest cavalry generals.

CHAPTER IV

INDEPENDENT RECONNAISSANCE

There is no part of military duty, save the art of chief command, which demands a higher qualification than the conduct of independent reconnaissance. For the paths which have to be followed in war by most subordinates are made smooth by orders, by training, by tradition, or by uniformity; their duty can be practised in peace with some approach to reality; the experience of their predecessors in war has been amply recorded. Of all these normal aids to excellence in war, there is only one—training—which will assist the aspirant to skill in independent reconnaissance, in anything like the same degree as it will benefit those whose duties are confined to the orthodox operations of war. To the independent scout or patrol leader, orders are hampering, tradition, useful tradition at least, has been lost, conformity is impossible, practice on manoeuvres is apt to be misleading, and the meagre records of previous war experience have been so overlaid with romance and exaggeration that many of them are useless. On the value of training, and especially self-training,[95] he may count, and general knowledge of the military art is essential to him; but for further aids to success, he must rely on resolution, on keen perception, and on quick understanding. Without resolution he will seldom be in a position to find out anything; without perception he may fail to

---

95. As Henderson pointed out, self-training is an important aspect of being successful in any endeavour.

find that which lies within his view; without understanding he may not grasp the meaning of that which he has found.[96]

In war the action of the most insignificant subordinate may turn the issue of a campaign. A battle may be won by the gallantry of a corporal who rallies a handful of men; the fate of Empires may hang on the drowsiness of a sentry or the shaken nerves of a private soldier ripe for panic. But of all subordinates, the scout is he on whose success or failure great issues are most likely to depend. The pursuit of information is so uncertain, so full of chances, that it may well be given to a couple of scouts to achieve the end for which a cavalry brigade is vainly striving. The possibility of rendering such service, of acquiring information of vital importance, should be always in the mind of the leader of an independent reconnaissance. He must never forget that his success or failure may mean the success or failure of the army to which he belongs.

The chief characteristic of independent reconnaissance is its uncertainty. A sufficient force well led[97] and employing contact methods, is almost certain to obtain some valuable information; protective reconnaissance can ensure the security of an army; but the leader of an independent reconnaissance can never be sure that his expedition, however skilfully conducted, will not be futile. In the first place, he may have been given an unreasonable task, such as the penetration of an enemy's screen before it has been broken, or the passage of a tract of country swarming with hostile inhabitants. Commanders who have never studied reconnaissance have sometimes exaggerated views as to its

96. And here, the importance of self-training is evident. Those who pursue study are more likely to be able to understand, where the undertrained person will struggle. On a battlefield, struggling to understand is likely to lead to death and defeat.

97. Leadership is a recurring theme, so its importance should be noted.

possibilities, especially with regard to independent methods; they consider that the result which has been achieved under the most favourable circumstances, should be possible under any circumstances, and as a consequence, patrols are frequently sent on impossible missions.[98] There is, secondly, the consideration that independent reconnaissance depends in great degree for its success on the faults of the enemy's protective system, and it cannot be expected that the enemy will always be obliging in this respect.[99] There will always be faults, no doubt, but sometimes they are not easy to find; and until the weak spots are discovered, the proportion of abortive efforts in independent reconnaissance is likely to be high. Lastly, there is the fortune of war; some fortuitous accident may spoil the most promising enterprise or block the most skilful design. For any of these reasons, an independent reconnaissance may fail through no fault of its leader; on the other hand, he may find his task well chosen, the enemy negligent, and luck on his side. This combination of favourable circumstances is not so rare as might be supposed; but the advantage it gives can only be secured if the right man is there to take it. To the skilful scout, in such case, anything is possible, but opportunity is wasted on a blunderer. When an independent reconnaissance is ordered, certain points should be made clear to the leader.[100] First of all is the information or the

---

98. Perhaps this is due to poor operational intelligence, poor planning and/or poor leadership. Nonetheless, Henderson's advice should be considered.

99. Countersurveillance, which is part of the discipline of counterintelligence, is seen as secondary to other operational functions. But here, Henderson underscores how being diverted to those "other" tasks can be a force's undoing.

100. Using the military "five paragraph field order," known by its acronym, SMEAC: situation, mission, execution, administration/logistics, and command/signals.

nature of the information that is wanted. This may be quite definite, or it may be general; and if information of both kinds or two separate items are wanted, one or the other should usually be made the more important. Thus, an officer who is told to endeavour to "find out whether the enemy is holding such a place, and whether his main cavalry force is on his right," should also be told which of these is to be the main objective of his quest, for he may well find it impossible to achieve both.[101] Then there is the direction in which he should look for his information, in order that he may not duplicate the work of others. This should always be secondary to the information itself. For example, such an instruction to a patrol leader as: "Go to Basingstoke if you can, and find out whether the enemy's reserve is there," is too strict, and limits both the initiative and the independence of the patrol to an unnecessary degree. For the patrol may reach Basingstoke and find no enemy, and return with its report, and yet this technical success may be of no use. A better instruction would be: "It is important to locate the enemy's reserve, which is possibly at Basingstoke. Try to find it in that neighbourhood. Other patrols are being sent to Hook and Oakley." The objective of the patrol leader is then clear. It is not to get to Basingstoke, but to find the enemy's reserve; and his mission is not successful until he has found it, if it is within reach. The more particular and definite the nature of the information required, the more general should be the instructions as to the route to be followed. Such available information as will assist the scout or patrol leader in the selection of his route should of course be given, but the actual choice should be left to the responsible man. The highest type of independent

---

101. A guide for asking good questions is Charles Vandepeer, *Asking Good Questions: A Practical Guide* (South Australia: Freshwater Publishing, 2017).

reconnaissance is that in which the leader is told what information is required and is given a free hand as to the steps he will take to obtain it. The excellence of Wellington's information in the Peninsula was due, in great measure, to his wisdom in refraining from hampering his intelligence officers with definite instructions; he knew his men and trusted them, and they, being left free to exercise all their ingenuity and skill, seldom failed in their missions.[102] If the information required is quite general in its nature, as in the case of several patrols being sent to discover the line of retreat of a defeated enemy with whom touch has been lost, the route to be followed by each patrol may sometimes be definitely laid down; but such reconnaissance can hardly be called independent. It is rather contact reconnaissance with insufficient means, and however skilfully the patrols may be handled, more risk will be run and probably less information will be obtained than would be the case if the reconnaissance were properly organised on either contact or independent lines. For the routes which are selected will probably be those by which the enemy is believed to have retired, so that the patrols, if they find the enemy at all, will make contact on his rear guard, which is exactly where he will be looking out for them. If the patrols are supported, as in a contact reconnaissance, the preparedness of the enemy is not a serious matter; but for an independent patrol it is infinitely more advantageous to approach the enemy from an unexpected direction—in this case, from a flank.

---

102. Moreover, the number of requests should be kept to a minimum to avoid a flood of information that will make analysis difficult, especially in an operational situation. Equally, a single but sweeping request for information can also be counterproductive. See Henry Prunckun, *Special Access Required: A Practitioner's Guide to Law Enforcement Intelligence Literature* (Metuchen, NJ: Scarecrow Press, 1990), 18–19.

In any campaign, except perhaps against savages,[103] or among an actively hostile population, there are but few intervals when independent reconnaissance cannot be usefully undertaken. However efficiently the contact and protective duties of an army may be performed, there is yet room for the employment of independent methods. A force employed on contact reconnaissance is practically forced from time to time to abandon its reconnaissance formation, and to concentrate, or at least to manoeuvre, with a view to tactical action. Such an operation is either the effect or the cause of similar changes of formation on the part of the enemy's mounted troops; contact reconnaissance on both sides is, for the time, abandoned, and the opposing forces of cavalry devote their whole attention to the combat. This is the opportunity for independent reconnaissance, for the small patrol to evade the hostile masses and probe the enemy's protective screen. It may chance that the enemy has not preserved the distinction between contact and protective troops; in such case, the resolute scout may well hope to penetrate until he touches something solid, foot or guns, welcome material for his report. To take advantage of such opportunities, independent patrols should always be in readiness, accompanying contact troops, pushing out from protective troops, edging out from troops in rear to hover on the enemy's flanks. Instructions for parties of this kind may be quite general; all have at heart the same object—to get within observing distance of the enemy's forces, and to bring back information about them.

When a force engaged on contact reconnaissance is only feeling for the enemy, and has not yet met with any opposition, independent patrols may, with advantage, be sent out to precede

---

103. This term is stated in the regrettable parlance of the time to describe combatants who should have been referred to as *indigenous*. See footnote 79 for a historical discussion of this issue.

it. Under favourable conditions these may be able to push far ahead, and to discover whether the main reconnaissance has anything tangible in front of it or is striking in the air. Such patrols are especially necessary when following up a defeated enemy; for contact troops in pursuit are frequently despatched in a wrong direction. The contact reconnaissance of the Germans after Spicheren was guided almost entirely by information sent in by independent patrols; indeed, so indispensable was this information, that it is an interesting speculation to consider the difference that a few well conducted patrols on June 17th, 1815, might have made to Grouchy and to history.[104]

It is, however, probable that independent patrols sent out from a contact reconnaissance force will find more difficulty in evading the enemy than will those despatched from other bodies of troops, for the reason that the mere presence of a contact force is likely to cause the enemy to bring his own mounted troops to that neighbourhood, even to the extent of denuding other districts. Therefore, when any large force is despatched on a contact enterprise, independent reconnaissance should become active elsewhere, to take advantage of the almost inevitable concentration of the enemy. For although the hostile cavalry may not be intended to act as a screen but may be properly employed as a contact or tactical force, yet its presence is a danger and a serious difficulty to small independent patrols. Large bodies of cavalry are, or ought to be, surrounded by a

---

104. History shows that because Grouchy was unable to close with the Prussians, he followed them according to the route specified in his operational orders. This was despite him hearing Napoléon's troops engaging with other sections of the Prussian and British–Dutch armies nearby at Waterloo. Although Grouchy eventually won his engagement with the III Prussian Corps in the Battle of Wavre (18–19 June), he was too late to help in the more decisive battle at Waterloo.

cloud of scouts and patrols for purposes of their own immediate information and security, and it is, therefore difficult to pass anywhere near them without detection. And as the masses of cavalry on each side will, as a rule, automatically draw towards each other, it is evident that the districts in which they concentrate will be those least favourable to independent reconnaissance on a small scale. It is therefore probable that information of urgent importance to the leader of a contact reconnaissance may be obtained by means of independent patrols which report, not to him, but to other commanders, possibly far distant. In the days before the introduction of the field telegraph and telephone, this was a decided disadvantage; but in an army in which modern methods of communication are properly established, it should be possible to keep the commander of the contact force supplied with all information which may concern him, no matter by whom it has been obtained.[105]

There is one advantage frequently obtained by successful independent reconnaissance on a small scale, which can hardly be hoped for if any other method be employed—that is, the ignorance of the opposing commander as to the information obtained. For if he becomes aware that his dispositions have been observed, it is quite possible that he may consider it advisable to alter both the dispositions and the plan for the furtherance of which they were designed. If he should do so, then the information may become for the time misleading. Troops employed on contact reconnaissance, or on independent

---

105. More than a century has passed since field commanders viewed field telegraphs and field telephones as "modern methods of communications." At the time of this writing, commanders have access to voice and data communications as well as live video of the area of interest that can be transmitted from reconnaissance troops as well as overhead surveillance vehicles.

reconnaissance on a large scale, cannot well accomplish their object without informing the enemy of their success; but it is possible, and indeed probable, that a small independent party, when it is so fortunate as to obtain good information, will do so unknown to the enemy, who may therefore still believe that his secrets are inviolate. The possible importance of such a situation is evident. As saith Vegetius: "There are no counsels and resolutions better than those which the enemy knoweth not of before you put them in practice. And therefore, when you come to know that your design is discovered to the enemies, you ought to change the orders."[G, 106]

There has been, in some campaigns, a tendency to overdo independent reconnaissance, and to despatch so many patrols that they get in each other's way and multiply the risks with no corresponding advantage. The officers of the general staff, who are responsible for the acquisition of information,[107] are the proper persons to initiate such enterprises; they should know at any time what information has already been obtained and what is still required and should therefore be able to select profitable and avoid futile objectives for reconnaissance.[108] Also, if the intelligence service is properly organised, these officers should be kept informed, by the general staff at headquarters, of the lines on which reconnaissance is being generally conducted, so that each may be able to appreciate the direction in which his efforts are likely to be most useful and in which there is least probability of his interfering with, or duplicating, the work of others. Circumstances will no doubt arise from time to time

---

106. Publius Flavius Vegetius Renatus, *De re militari* (abt. 450).

107. In this case, Henderson is discussing an operational level intelligence unit.

108. Such requests for information are termed *intelligence requirements* (IR) or *intelligence collection requirements* (ICR).

which will justify subordinate commanders in initiating independent reconnaissance, but as a rule they should use their discretion to do so sparingly. If not, there will assuredly be confusion and wasted effort, as there was after Spicheren, when German patrols sent out by army corps and divisional cavalry, by army corps and divisional headquarters, by army headquarters and by cavalry divisions, were for two days crossing and recrossing each other, without any proper instruction, direction, or objective, discovering each other very often and the French very seldom. This occurred under most favourable circumstances, when there was practically no opposition offered; and to this fact it is due that the reconnaissance was wasteful in effort only and not in life. Some isolated items of information were obtained, and these more by accident than by design; but it was not until a general staff officer, to whom all sources of information were open, himself steered a patrol through the meandering horsemen, that the main French army was located. Had the intelligence organisation of the Germans been adequate to the requirements of a great campaign, there need have been no confusion and very little misdirection; it is improbable, in fact, that touch would have been lost at all. It may of course be argued that this was a case for the employment of contact and not of independent reconnaissance, and so indeed it was; but it must be remembered that the cavalry divisions, which actually were idle, might possibly have been obliged to concentrate for tactical operations, thus giving an opportunity to, and throwing the responsibility on, the independent patrols. And, in any case, whether it is or is not likely that independent reconnaissance will achieve great results, those who are employed on it should be given every assistance and every chance of success, that men and horses may not be wasted in useless effort.

CHAPTER V

THE SCOUT

It is often asserted that a good scout is the product of natural aptitude and of environment, and that training can do but little to improve men who are by nature fit, while it is wasted on those who are not specially gifted. This opinion, which is confined to those who have had no experience in reconnaissance, is misleading, for it implies that there is some peculiar virtue or natural gift possessed by the born scout, without which success is impossible, whereas it is by a combination of ordinary military qualities that success is achieved. There is, no doubt, one attribute which is indispensable—that is, courage: there is no hope of making a scout out of any man who is not resolute. But given courage, there is no other natural gift the lack of which renders a physically and mentally sound man unfitted for scouting. The fact is that the qualities which in highly developed form are required in the scout, are exactly those which are required in the soldier, and they are mostly qualities which can be acquired by training unconscious or deliberate. The average sound mind can be unconsciously trained to alertness and, caution by a life of danger or a struggle for existence, to fortitude by hardship, to swift decision by early responsibility; so also, may the sound body be unconsciously trained to suppleness, endurance, and accuracy. The man thus equipped is material ready to be fashioned into a scout; he is the natural "guerrillero,"[109] the soldier for petty war,[110] the partisan.[111] To

---

109. Spanish for *guerrilla*.

make him a finished scout, he must be taught something of the art of war, and this is sometimes a difficult matter. On the other hand, the trained soldier, officer or man, begins with a knowledge of civilised war; but his faculties must be deliberately trained to the standard necessary for successful service in the difficult and dangerous duty of scouting. It is not easy to say which is the better type. For different natures of warfare one or the other may be preferred: for one enterprise the educated officer would be chosen, for another the acute and skilful partisan. The partisan type, with highly developed instincts, is often lacking in the general knowledge acquired by education;[112] and in modern warfare on a large scale, the scout who lacks general knowledge will certainly miss valuable information. The other type, whose procedure is based on reason and calculation, is frequently at a loss from want of experience and practice. The first knows how to find, the second knows what to look for. In rare cases the highest development of both types is attained by one individual: such men make history. Waters and Grant in the Peninsula, Hodson in India, Ashby in America, directly influenced the course of campaigns. The commander is indeed fortunate who can find in his army one man of this calibre. Although few can hope to rival the skill of such masters of the art of individual reconnaissances, yet it is undoubted that by earnest study and endeavour the average man can attain a high degree of efficiency. Even a short course of training under a competent instructor will affect a marked improvement in the skill of the ordinary intelligent soldier.[113] The eye may be trained to swift

---

111. A *partisan* is a person who is an irregular trooper whose group harasses the enemy through acts of hit-and-run fighting and/or sabotage.

112. Here, Henderson uses the term *education* to mean *training*.

113. Henderson is correct in his assertion. The subject literature on training and personnel development shows both effectiveness and
–continued–

and definite observation, the mind to correct deduction, the memory to retentive grasp; the body may be trained to move silently and to lie close. There are also certain principles of the art of scouting, drawn from the experience of others, which may be taught.

The scout has ever been a favourite hero of romance, but the qualities in a scout which appeal to the popular fancy are not exactly those which are in reality most important. The public, who look at all military affairs solely from the point of view of sentiment, regard a scout as one who is constantly in grave danger, from which he escapes (when he does escape) by the exercise of peculiar adroitness or by good fortune. In fact, they measure the fame of the scout by the amount of danger he escapes;[114] and it is frequently forgotten that the scout's business is to acquire information, and that his reputation should depend on his success in that line, not on his heroic adventures. The scout who merely gets into danger and out again achieves nothing. He proves only that he is fitted for the first part of his duty—that is, to take care of himself in the immediate presence of the enemy. It may be admitted that this first stage—the means to an end—is the fascinating part of scouting, and it is this which in the popular mind has obscured the final stage—the end to be attained—namely, the acquisition of information. Every scout must be prepared to take risks and should be trained in such a way that his chances of surviving the necessary dangers may be increased; but the best scout is he who attains his object while exposing himself as little as possible to danger. Romance

---

efficiency of the individual improves with the amount of training received. Even a small amount of training will show improvements.

114. As a contemporary illustration of the derring-do scout, see Larry Alexander, *Shadows in the Jungle: The Alamo Scouts Behind Japanese Lines in World War II* (Boston: Dutton Caliber, 2010).

demands hairbreadth escapes and perils surmounted; but the general in the field wants information and takes more interest in the dry narrative of ascertained facts than in the most thrilling details of unfruitful endeavour.[115] Therefore, he who aspires to be a successful scout must never lose sight of his ultimate object—information; and however, enthralling he may find the pursuit of his object, he must remember that exciting adventures usually mean difficulties and delays, and should be avoided when possible. The procedure of a successful scout is guided by a combination of courage and caution: both these qualities are essential, and the lack of either is fatal to efficiency. A timid scout is not of much use, but a dead scout is of no use at all, and a scout who is captured may be a danger.[116]

It is therefore desirable to divide the qualifications of a scout into two classes and his education into two parts: the first dealing with the preservation of his life and liberty in difficult circumstances, the second with the acquisition and transmission of information. The distinction may be made quite clear if the case be imagined of an officer who knows nothing of reconnaissance being led to a post of observation by a skilful poacher who knows nothing of war. The working of the combination would resemble the operations of a trained scout: the conflicting claims of the desire for information, and the

---

115. This advice cannot be emphasised enough. The art of report writing has a foundation in fact and logic. Creative writing is based on plot, crisis/tension, characterization, dialogue, setting/context, as well as the point of view of the narrator, and the use of literary and rhetorical devices.

116. I would argue a captured scout *is* a danger because, through interrogation, the opposition can determine his/her commander's plans. This can be done by applying reverse-thinking to the scout's instructions about what information was being collected and the urgency for it to be obtained.

difficulty of obtaining it, would have to be argued out between the two men in the same way as these claims must be balanced in the mind of the scout; the caution of the guide and the eagerness of the observer would in turn influence their united action. Moreover, if it were desired to turn this pair into efficient single-handed scouts, their education would have to proceed on quite divergent lines, the soldier being taught to take care of himself and the poacher to know what to look for and how to report it.

The final step of a scout's education, and this is also the final test of his capacity, is practice in war; and unless the tyro is fitted for his task by preparation of some kind, his career is likely to be either short or inglorious. Assuming, however, that he has the necessary qualities to give him a fair start, he will find the education severe and the test searching. There is no occupation that will more quickly disclose a weak string in the nerves or a white spot in the heart. The courage of the efficient scout must be of a fine temper: mere hot-headed, blundering bravery, useful enough behind a bayonet, is out of place on reconnaissance: its exercise will lead only to disaster. He should be of equable temperament, his coolness undisturbed in danger, his resolution unmoved by difficulties or obstacles. He must be contented to play the game for its own sake; it is likely enough that his gallantry will be unseen by others, and unrecorded.[117]

If scouts were required to work invariably single-handed, there can be no doubt that the number of those who fail from loss of nerve would be much greater. There are a few men who prefer, and excel in, solitary reconnaissance; but as a rule, the assistance of a companion adds to the attraction and eases the tension of scout-work. Not only is this the case, but the results

---

117. And this is the case with all intelligence work. It is rare that an intelligence office has his or her exploits retold in books or cinema.

are usually better when scouts work in pairs; companionship gives confidence, and observation is cool and deliberate. It may be laid down as a principle that in ordinary circumstances and with average men, two scouts are more valuable if working together than if allotted to separate duties. There are, of course, occasions when a scout must work alone; the circumstances which dictate such a course are usually those in which concealment is both difficult and necessary. For the actual penetration of an enemy's outpost line, or the close reconnaissance of a vigilant post, the single observer may be the only possibility. Such enterprises, however, are exceptional: for general work, men should not be alone.

Although, however, it is usually inadvisable for scouts to work alone, it is convenient, in discussing the various points to which the attention of scouts should be directed, to consider the scout as working alone and independently. The suggestions put forward for the guidance of a single man can easily be applied to the procedure of two or more.

The precautions which may be observed by a scout in approaching and observing an enemy are many, but so varied are the circumstances under which his difficult duty must be carried on that there is almost no precaution which may not, on occasion, be properly neglected. In every case he must balance chances; he is of no use if dead or a prisoner, and yet the information he is in search of, may be of urgent importance. One of the first rules is concealment: to see without being seen. Another is, that when concealment is impossible, it is injudicious to get within close range of the enemy. These would seem to be axioms; yet both have been disregarded with impunity and success, and of this an instance may be given.

During the South African War, two scouts were instructed to report on the condition of a "drift" or ford on a road by which the enemy's forces had retreated. The scouts succeeded in reaching unobserved a hill about three hundred yards from the drift, which lay on their left front, the intervening ground being quite open. Near the drift, but farther to the left, was another rising ground, from which a covered retreat seemed possible, and just across the river was a rocky kopje.[118] Watching this kopje carefully, the scouts came to the conclusion that it was occupied by the enemy. There were then three courses open to them, and these they discussed. First, to retire, in which course, no doubt, they would have been justified; second, to make a detour by the rear and try to reach the second hill, from which something of the drift might be seen; third, to make for the drift and take all risks. The information was urgently required, and the men knew their enemy. "If they think we don't know they are there," they agreed, "they will wait for us to cross the drift and won't shoot while we are on this side." So, they mounted their horses and rode at a walk towards the drift, laughing and talking. They rode into the stream, where they were hidden by the high banks, and having examined the drift, bolted out at a gallop and made for the cover of the second hill. They were well on their way before the enemy opened fire. They reached cover untouched, and although followed for some distance, returned in safety to camp.

This incident may serve to illustrate several principles of scouting. The factors which contributed to the success of the scouts may be stated in order; they are all important: concealment in advancing; the occupation of a good post of observation; recognition of a possible danger; prolonged scrutiny

---

118. A *kopje* is a small hill on the African grassland.

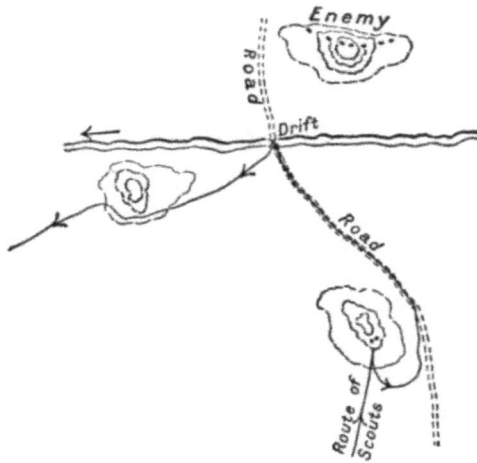

of the dangerous point, and detection of the enemy; consideration of possible courses, and the balancing of the value of the information against the risk of obtaining it; knowledge of the enemy's methods; and lastly, a definite plan. These may all be considered general rules of scouting; and it will be seen that these scouts, when they abandoned the principles of concealment and of avoidance of the enemy on open ground, brought into play a stratagem based on their knowledge of the enemy, and suited to the particular case to which the general rule was not applicable. This precaution justified the departure from recognised methods.

The art of concealment may be acquired, to some extent, by the use of ordinary common-sense; by avoidance of crest-lines, summits, or open ground; by caution when moving and quiescence when stationary; and above all, by taking advantage of darkness. To these the trained scout adds a knowledge of the effect of different backgrounds, and of the visibility of objects in sunlight and shadow. This knowledge can only be obtained by close observation, aided, possibly, by instruction. Important also is the power of controlling the body to absolute stillness ("freezing," as it is called by hunters)—a power which is a natural endowment of most wild animals, but which cannot be acquired by man without practice. The value of this faculty is due to the fact that signs of life and signs of motion are to the untrained eye synonymous, and that to detect the presence of life ninety-nine people in a hundred try only to detect motion. The idea of perfect stillness is dissociated from the idea of the presence of living beings; a man may consequently remain unobserved, if only he does not move, in a position where concealment seems impossible. Closely allied to this power of stillness is the ability to move so slowly and steadily that the motion is difficult to detect. This also can be practiced, and by many can be acquired without much difficulty. Patience is one of the attributes of a good scout, and the patience which may be required in many trying situations may also be improved by this practice in controlling the muscles and nerves.

The intimate connection between movement and visibility suggests some other precautions. Thus, a scout who finds it necessary to place himself in the possible range of vision of the enemy should have his movements well under control. For instance, a mounted scout approaching a position suitable for observation, such as the crest of a hill or the edge of a wood, should not attempt to look out from the saddle, however carefully he may be able to conceal his horse; for the movements of the horse can never be absolutely under control, and any quick motion may mean detection. The choice of a horse should be a matter of some care also; a nervous or fidgety animal is a source of danger.

Another point is, that full use should always be made of a good post of observation before moving off in search of a better; for it is no uncommon experience for a scout to lie for an hour undetected in one spot, and to be chased five minutes after he has left it.

The concealment afforded by the cover of darkness is of inestimable value. Not only to approach the enemy, but often to escape him, the dark hours may be utilised. It is generally a good principle, when the hostile force is near, and must be observed daily, for the scout to make his approach to the enemy by night, so as to reach his farthest point of observation before dawn. He may thus be posted in concealment ready to make use of the first light for his observation; he may very probably obtain all the information he requires without finding it necessary to move from the post he has originally chosen, and he should have ample warning of any hostile movement which might threaten his safety, or his retreat. For the scout whose mission it is to penetrate the enemy's lines, the help of darkness is even more necessary; for it is only on very rare occasions that the passage of the enemy's outpost line can be achieved in daylight. He who successfully penetrates the enemy's lines, also must, as a rule, pass them again on his return; and the old-fashioned method of breaking out of the hostile camp by risking a dash in the open, after the manner of the intrepid Colquhoun Grant[119] at Huerta, presents but small chance of success if attempted under the fire of modern rifles. The return journey also should be accomplished by night if possible.

Skill in night operations, like most attainments of the good scout, can be acquired by practice. The principal points for study are the sense of direction and the judgment of time and space. For the first there are many guides—compass, stars, wind, running water, hills, roads, and fences.[120] Some guide of this

---

119. A reference to Lieutenant-Colonel Colquhoun Grant (1780–1829). He was an intelligence officer with the British Army during the Napoleonic Wars.

120. And, of course at the time of this writing, GPS—global positioning system.

kind is usually necessary, without one, movement in darkness is apt to degenerate into mere groping. Of course, like Lord Hawke at Quiberon Bay, who "took the foe for pilot, and the cannon's glare for light," scouts have been known to achieve a successful reconnaissance by night, guided only by a returning patrol of the enemy; but convenient assistance of this kind cannot always be counted on. A scout should seldom undertake a night enterprise in presence of the enemy, unless he has a definite idea of where he wants to go and how he means to get there.[121] Having decided these questions, the skilled man should have little difficulty in finding his direction, either by moving on a chosen line, which may be fixed by star, compass, or wind, or by utilising natural landmarks as signposts on his way.

The judgment of time and space by night is almost as important as the accuracy of direction and is quite as difficult. The beginner, indeed, usually finds it much more difficult. The ordinary educated man can make a fair guess at correct direction but will experience an uncomfortable vagueness as to the distance he has travelled and—if he be not allowed to consult a watch—as to the time. In the estimation of time, of course, the stars or moon may be of assistance to those who have been taught to observe their movements; and on many occasions there may be no danger in showing enough light to read a watch; but any scout who works much in darkness will often find it necessary to trust to simple calculation of the progress of the hours. Time is of importance for two principal reasons: the first, that the hours of darkness are limited, and to be exposed by daylight, while still engaged on an operation which can only be safely carried out by night, puts the scout in an uncomfortable situation. The second point is that time is a useful check on

---

121. Compare that with today's technological aids—night vision goggles and GPS navigational devices.

distance traversed, and the estimation of distance is the chief difficulty to overcome.

There is no royal road to the acquisition of this power of estimating time correctly by night. Practice and observation are the only teachers, and their best pupils can attain only to approximation, never to certainty.[122] But the measure by which the trained man falls short of certainty is as nothing compared to the difference between his approximation and the vague attempts of the beginner.

A continued effort of attention is required to obtain good results in the matter of estimation of distance traversed. The safest system is, probably, to estimate, successively, short distances as they are traversed, keeping the total distance in the memory.[123] The respective extent of these separate distances depends chiefly on the nature of the ground; in difficult country they must be shorter than in open ground, and in any case, an estimate should not be carried over any obstacle which causes delay or divergence. A necessary halt should always be made the terminating point of one estimate and the beginning of another. Time has been noted as a useful check on distance, and its value depends on knowledge of pace. Estimates of distance traversed by night by means of calculations of time and pace, are, however, unsatisfactory for distances under three or four miles, and their accuracy is much influenced by delays. Nevertheless, in many cases the time check will be found useful, especially if the journey be long and the course moderately straight. Fair

---

122. Hence the title Henderson gave to his book—*The Art of Reconnaissance*.

123. Now scouts have devices to help in this regard—pedometers, GPS route trackers, and live transmission of these data via satellite to command headquarters for plotting and analysis.

accuracy in judgment of pace in darkness is, with practice, not difficult to acquire.

The conditions most favourable to night scouting are a dark night, a clear sky, and a knowledge of the existence of certain recognisable landmarks. The darkness gives cover, the clear sky gives guidance, and the landmarks make position from time to time certain. In choosing suitable landmarks, either from view by day, from maps, or from description, some should be, if possible, continuous lines such as roads, streams, or ridges, and these may be valuable either parallel to, or across the line of route. The value of transverse lines lies in the fact that there is a practical certainty of hitting them off at some point in their course; and should the scout have missed his direct route, he will usually know to which side of the point he was aiming at he has diverged, and by following the line of his landmark will probably strike his direct route again. Even should he fail in this, the transverse line gives him his distance, and this is always useful.

The following account of a despatch riding expedition, related by an officer who was employed on intelligence duty during the South African War, illustrates the possibility of finding the way at night by a combination of different methods. A large force, moving across the veld[124] to take part in a combined operation, was approaching the district in which the main body was known to be operating. The commander of the moving force was, of course, very desirous of establishing communication as soon as possible, and for this purpose asked the intelligence officer to endeavour to reach the other force and deliver a message of urgent importance. The officer having selected a scout to accompany him—a man who in peace time had followed the occupation of a transport rider and had a

---

124. An African grassland used for grazing livestock.

general knowledge of the country—started on his errand, accompanying, at first, a cavalry force which was ordered out with the object of opening up signalling communication. The cavalry was delayed by some skirmishing, so the intelligence officer, choosing a favourable opportunity, pushed on with his scout, and eluding the enemy, made his way at speed, towards the point where the main body was supposed to be. After a somewhat circuitous journey, he succeeded in delivering his message, and at nightfall prepared to return to his own force.

Before starting, he consulted with the scout as to the route. The distance to be transverse was about eighteen miles, and from his indifferent map, the officer was able to assure himself that his camp lay nearly due south. The scout professed his readiness to follow, by night, the roundabout course which they had pursued by day, but was doubtful of his ability to find the camp, with certainty, otherwise. The officer was confident that he could keep direction fairly by the stars but could not promise absolute exactness. In consideration of the possibility that they might again have to deviate, to avoid the enemy, they agreed that the stars offered the most certain guide, and that by steering a direct course, they would not only shorten the journey, but would avoid the delay caused by tracing out their previous course by the help of landmarks. But in order to make certain of not overshooting their mark, they decided to edge off in a direction (to the right, as it happened) which would tend to bring them across their former course at some point, which they hoped, even in darkness, to be able to recognise. They moved, therefore, as nearly as possible due south until they estimated that ten miles had been covered, and then changed course to a direction about south-by-west. Just as they were beginning to have doubts about the accuracy of their navigation, a dog barked in the distance.

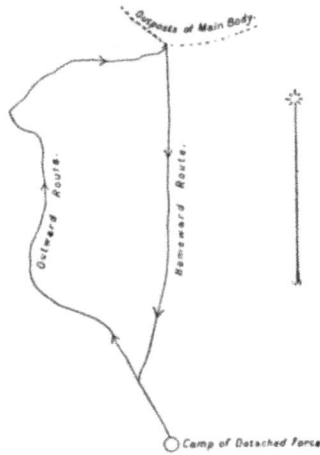

The scout stopped and listened to get the direction of the sound, and then said: "That dog might be at a farm I saw this morning. If it is, then the wagon road we started on ought to be close by." They turned due west and found the road within a hundred yards. The scout dismounted and walked along the road for a furlong[125] or so, the officer following with the horses. Then the scout announced that they were three miles from camp. "I remember this small wash-out," he said, "and the bare rock showing, and the wheel track of that Boer wagon that left the road to clear it." In half-an-hour they were within the outposts.

This method of edging off the direct course, in order to make certain of hitting off a continuous landmark, is often a useful precaution.

It is necessary for a scout to learn to move silently, and although at night the ordinary precautions required are not very

---

125. A furlong is equal to one-eighth of a mile, or 660 feet (i.e., 220 yards, or approximately 201 meters).

dissimilar to those suitable for daylight, yet some extra care is necessary. When on foot and off a road noise is most likely to be caused by stumbling, owing to unseen obstacles on the ground, or unevenness of the ground itself. Stumbling may be avoided, to some extent, by moving with both knees bent and the balance of the body rather thrown backward, keeping the weight on the rear foot until the other is safely planted. If so close to the enemy that absolute silence is necessary, the only safe method is to move on hands and knees, carefully removing any dry sticks which might crack under the weight of the body. If the scout be mounted, or leading his horse, then silence cannot be ensured; but unless an attempt is being made to penetrate an insufficient or badly-posted outpost line, it is better to leave the horse well in rear. If, however, a horse must be taken, he will usually make less noise if led than if ridden. Where good roads exist, the bicycle is likely to be of value in night reconnaissance, partly because of the great distances which can be covered, and partly because it affords silent and speedy means of escape. This method of locomotion is, however, too swift for a close approach to the enemy; but a bicycle may be led with safety to fairly close quarters.

There is no branch of a scout's education in which earnest practice is more speedily and certainly repaid than in this matter of movement by night. If any should doubt this assertion, a month's trial will usually serve to convince them. There is not much difficulty in learning how to find the way about a country by day. Every scout should be able to read a map and should understand the use of a compass; he should be taught always to observe landmarks, and to remember the general direction of the route he is following. All these details can be practised in peace time, and practice is the best method of acquiring the sense of locality.

To take advantage in daylight of the concealment afforded by natural objects is, of course, an elementary precaution, and any sensible man will utilise obvious cover such as woods, hills, rocks, or buildings. It requires some discernment, however, to recognise all the possibilities of cover, and it is astonishing to observe how insignificant are the irregularities of ground or herbage which may serve for the concealment of a skilful scout. Obvious cover may be sufficient for a fixed observer, but it is seldom continuous, and therefore cannot give protection when movement is necessary. It is in moving from cover to cover, from wood to building, from hill to stream, that the beginner finds difficulty. He must learn, or be taught, to appreciate the value of shadow or broken sunlight, of small hillocks or depressions, of bushes or heather, or even tufts of grass. He should be shown how a protection so thin that it merely blurs the outline may give effective concealment. Practice under observation is useful, that the learner may know wherein he failed, or at what points his presence was detected. And, of course, it must become habitual with him, before attempting a difficult passage, to consider carefully the probable position of the enemy, a condition which governs the whole problem.

Of the possibility of utilising cattle, horses, or other animals as cover, and especially if they are used to conceal movement, it is sufficient to say that care should be taken that the scout is accustomed to the habits of the species of animal he selects for this purpose, and that the animals should be tame and not likely to be disturbed by his proximity. For any unusual movement may draw attention, and thus defeat the scout's object. The movement of startled animals, whether wild or tame, has many times led to the discovery of a hidden observer. Wild birds, especially, are easily disturbed, and are a constant danger, as their presence is difficult to detect until too late, while their signs of alarm are unmistakable. Sheep and cattle also must, in unfrequented parts, be approached with caution by a stranger, and horses at pasture are often nervous.

With regard to the study and practice of concealment in scouting, it may be said, generally, that the quality of which the acquisition or improvement is of most importance is patience. Without patience it is difficult enough to learn how to hide; but even if a smattering of this art be learned, success in performance cannot be looked for unless impatience can be thoroughly controlled. Not only the natural impatience, which desires speedy success, a quick return for good work, but impatience caused by the slighting criticism of those who despise caution and mistake rashness for courage.[126] Rash scouting will certainly lead to heavy losses, which have a discouraging effect and react so forcibly on the survivors as to cause hesitation (a sentiment very different from caution), and in the end, moral inferiority.

Concealment has here been given first place among the principles of scouting, for the reason that reconnaissance which is carried out unobserved is not only safer for the scout but is usually more satisfactory in its results. By this method also, the enemy is not informed that he has been observed. So long, therefore, as success is possible by either overt or hidden operations, where, in fact, there is free choice without prejudice to the enterprise, the concealed method should be preferred. In certain cases, no doubt, wilful exposure may be the better course, as when the scout desires to verify his observation by drawing the enemy's fire, or ventures to bluff a timid opponent by a confident advance. But as a rule, the scout should try to work unobserved, provided always that this endeavour promises a reasonable possibility of the attainment of his object.

---

126. i.e., through maturity and experience.

There is, however, almost no imaginable combination of circumstances (save, perhaps, in forest fighting) in which the position of the enemy, or the natural or artificial cover available in his vicinity, would so favour a concealed reconnaissance as to enable a scout to carry out his duty successfully in daylight while taking no risks of exposure. In practically every reconnaissance, cover must be left at some point or points. Having once come under the possible view of the enemy, the best course for the scout to follow is to assume that he has been observed. He must endeavour to mislead a possible watcher, to detect a possible ambush, to evade a possible interceptor. The circumstances of the particular case will, of course, dictate his actions to these ends, but there are a few principles which are useful for general guidance.

One of the first of these is the avoidance of any appearance of apprehension. Even if a scout is certain that his movements are under observation of the enemy, he should try to appear heedless and unconscious of danger. For a civilised foe usually prefers to capture a scout rather than to kill him, not from any humanitarian sentiment, but because he is a possible source of information. There is therefore a tendency to allow a single man, or a small patrol, to approach to such close range that a quick interception or a sudden summons to surrender may affect a capture.

A scout who has reason to believe that the enemy is watching and awaiting him, should not by any sign betray his suspicion, but should manoeuvre for a favourable position, from which he may either verify his opinion, or avoid the danger. An example may serve to make this clear.

Let us suppose that a scout finds it necessary to advance across a succession of open parallel ridges and thinks it probable that on one of these he will find a patrol of the enemy. From the crest of the first he would reconnoitre the second for signs of his opponents. If unable to satisfy himself that the crest in his front is not occupied, he should assume that it is, and endeavour to out manoeuvre his possible adversary. Probably his best course now lies in following the rule "change direction when out of sight." His actual procedure might be to retire and mount, if he has a horse, cross the ridge openly and carelessly at a walk, and advance into the valley.

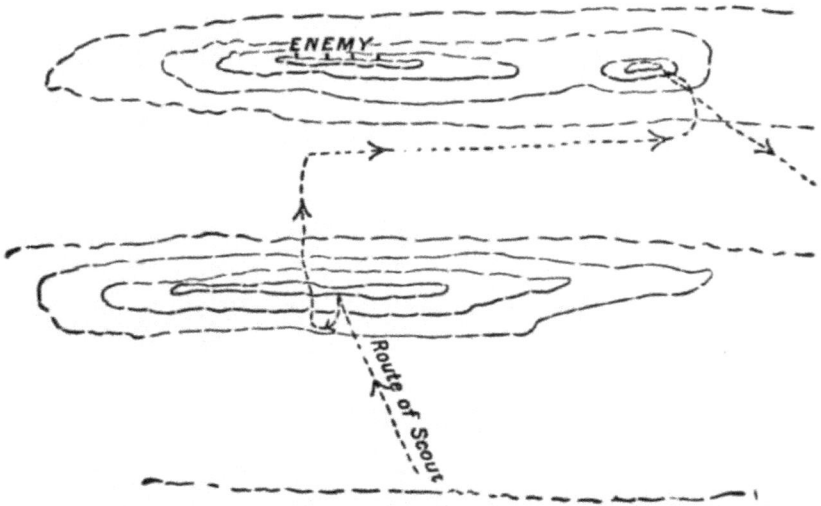

He should have been able, from his former position, to ascertain that no enemy was on the forward slope of the ridge and should feel reasonably certain that the bottom of the valley will be out of the range of vision of a concealed enemy on or behind the summit. When the scout reaches the valley, therefore, he should make a sharp turn and move at speed for some distance to the right or left, and if the ground be suitable, should make his way quickly to the crest at an unexpected point. If enemies are on the hill, they will probably be watching the point at which the scout would have appeared had he continued his original direction, and this diversion of their attention will put the scout in a favourable position.

This plan of changing direction when out of sight is one that may very well be used as a general precaution by a scout. The point at which he emerges from a wood, for instance, should never be in prolongation of his route before entering the wood. He should never, in fact, appear where he would naturally be expected. But in order to make himself expected, he must show no sign of suspicion. The advantage to be derived from unexpected movements is twofold—they give the scout an additional chance of detecting the enemy before he is himself seen, and they may induce an enemy, who is able only occasionally to observe them, to magnify a single scout into a patrol—an error which would be all in favour of the single scout. For not only might it lead to the retirement of the enemy, if merely a reconnoitring party; in any case, by widening the necessary field of observation of the watcher, it would increase the single man's chances of avoiding further notice.

If a scout, when unconcealed, should, through lack of control, betray by any sign or movement his sudden discovery of a hitherto concealed enemy, there is usually nothing to be done but to retire and try another route. For if within range, it is almost certain that he will be immediately fired on; and as he is at a disadvantage, he should go at once. Even if not fired on, he has given away one of his chief advantages—his supposed ignorance—and if he continues his advance on the same line, must expect to be made a target at the convenience of the enemy. It is, however, just possible that the concealed observer may be unsupported, and may also have been so foolish as to give away his advantages, either by allowing his opponent to come too close, or by choosing a post of observation where he cannot properly use his weapons (if he has climbed a tree, for instance), in which case it may be good policy to show fight. But it is seldom judicious for a scout to engage in combat unless he has to fight in order to escape. The use of firearms is almost certain to attract attention to his operations, and attention is just what he wants to avoid.

A consideration of the difficulties which attend the advance of a scout towards the enemy, leads to the conclusion that while moving he is at a disadvantage, compared with a concealed observer at the halt. This conclusion is borne out by those who have had experience of reconnaissance in the field; they will usually admit that there is an instinctive feeling of security when halted in a suitable spot, and of insecurity when on the move. This is due chiefly to the fact that a man can see better and is less likely to be seen when stationary than when moving. When a scout detects the enemy before he is himself seen, the discovery is almost invariably made while he is at the halt. While moving, his enlightenment frequently takes the form of a bullet, or a rush of mounted men to cut him off. The advance, therefore, should be made by stages; before leaving cover, he should scrutinise his intended line of advance, and if possible, should select his next halting-place. For an independent scout to neglect this precaution is simple foolishness; he is putting his life, and, what is more important, his mission, at the mercy of the enemy. The proportion of available time which should be spent in observation at the halt naturally depends on the conditions of the case; the nature of the reconnaissance, of the ground, of the enemy, of the weather, of a hundred varying factors, may affect the decision. The scout must judge when he has informed himself sufficiently of the dangers or aids which lie between him and his next proposed halting-place; his examination will be more detailed, and therefore more prolonged, in localities where he has reason to apprehend danger, as in the near neighbourhood of the enemy's position, than it need be in places to which the enemy does not usually penetrate. Also, if from any post of observation, he can gain information of military value, or of the kind he has been directed to procure, he is justified in delaying his further movements until he has learned all he can.

The security of a scout depends, in large measure, on his ability to recognise and to estimate possible dangers. He must see his enemy before he is himself seen, and to ensure this, he must try to divine the most likely positions for the enemy to be. In this endeavour, he may have the aid of information, either supplied by others or obtained as the result of his own observation, of experience in scouting, or of knowledge of the ways of the enemy. In order to guess the probable position of a possible enemy, a scout must try to look on the situation from the enemy's point of view. He should note the spots which his opponents might suitably occupy, either as outpost or as outlook positions. He should be always alert to note any sign of movement, especially near such places, for, as has been already noted, it is by signs of movement that the presence of life is most easily discerned. For a man with good eyesight, the best method is to examine the country with the naked eye, using field-glasses or telescope from time to time for the closer scrutiny of dangerous spots.[127] If his suspicion be definitely directed to some particular locality, he should make up his mind between two possible courses: either to make the doubtful point the immediate objective of his reconnaissance, and thus verify his suspicion, or to avoid it at a safe distance. In any middle course, as usual, there is unnecessary risk.

---

127. Scouts now have access to state-of-the-art optical surveillance devices, including digital equipment with powerful zoom, thermal, and infrared capabilities. Because these devices are digital, the data can be transmitted to commanders as events unfold.

The observation of tracks may often enable a scout to locate the enemy. A knowledge of tracking is therefore useful to a scout, and particularly the ability to distinguish between fresh and old tracks. The details by means of which this distinction may be drawn vary in different countries according to the nature of the ground and the climate, but are mostly connected with sharpness of outline, moisture, or wind. A track which is quite fresh has a sharp outline, which in time becomes blurred—in sand by the gradual falling of particles, in mud by the percolation of water. By study of the climatic conditions of the country in which operations are taking place, it is possible, by observing the degree to which the outline has become blurred, to estimate the time that has elapsed since the track was made. If a shower of rain has recently fallen, it is usually easy to see whether the track was made before or after the occurrence, by noticing whether rain has fallen on the tracks. It may be noted that the action of rain on tracks destroys their sharpness very rapidly. The action of wind also accelerates the blurring of tracks on sandy ground. A little practice, however, is worth a good deal of theory in the matter of estimating the freshness of a track, and also in learning to follow one. Tracks should be made from time to time close together, on all sorts of ground, and compared, and when actually on reconnaissance, the scout can compare tracks which he finds with his own. In this way, minute but important differences will become apparent. Attention should not be confined to the actual print of hoof or foot, but notice should be taken also of the displacement of grass or twigs, of horse droppings, and of the signs of occupation of a halting-place. In following a track, the scout should endeavour to distinguish it two or three yards ahead; it is difficult to pick up a track and easy to overrun it if the tracker keeps his gaze directly downwards.

When the scout, himself unseen, has detected the enemy, he must use his discretion as to his further action. It may be that this very discovery is the object of his reconnaissance, and in this event his success is nearly attained; he has only to ensure that his information reaches his employer. On the other hand, this particular enemy may be merely an obstacle in his path, and, if so, must be circumvented. This means, as a rule, the beginning of a new reconnaissance, under altered conditions, and a new plan is necessary. Or if the position or movements of the discovered enemy be such as to threaten immediate danger to the scout, measures for escape may have to be taken at once.

Knowledge of the usual methods of the enemy is of great importance in scouting, as in the other operations of war. This knowledge may be derived from intercourse in peace time, from study of previous wars in which the enemy has taken part, or, best of all, from daily experience in the early stages of the campaign. The peace habits of a nation, even of a regular army, are often strangely altered by war; their war methods also change from time to time; the knowledge, therefore, which is based on peace acquaintance or on historical study, is sometimes found to be ludicrously misleading in the field. But experience, gathered by close observation of the enemy's action and by deductions as to the motives for his action, is often a sound enough basis on which to form a plan to elude or outmanoeuvre him. The scout should always study his enemy: note the usual strength and range of his patrols or reconnaissances, the system of his outposts, the speed of his horses, the skill of his scouts. Any of these may be the decisive factor in the selection of a plan of reconnaissance: all of them are essential details for everyday use.

A scout while on reconnaissance should frequently consider both his immediate and his ultimate line of retreat or return. This does not mean that a direct line of communication to his rear is necessary; he is by no means bound to return along the line he chose for his advance. As a general precaution, in fact, a scout should not, if he can help it, return by the route of his advance:

snares are usually set in "runs." Rather should he make an early selection of alternative lines for his return; and as his reconnaissance progresses, he should consider at each stage which of these lines is for the moment most suitable, and how in emergency he would reach it, so that, if surprised, he may instinctively turn in the right direction, with a rough plan of escape ready in his mind. For the great danger of surprise is the hesitation it usually causes. If met confidently, surprise loses much of its effect.[128]

A scout, of course, must keep a lookout on his flanks and rear. An enemy interposing between him and his own force is an added danger, and the cases in which such interference will not necessitate a change of plan are but rare. If determined to penetrate the enemy's lines, interception may be regarded with equanimity; and if discovered in time, it is not of necessity alarming. But the passing of the hostile lines is not an everyday occurrence, and in performing the usual duty of observing the enemy from the front or flanks, there is a grave risk of discovery by any body of the enemy which, having gone beyond the scout, is returning to its own lines. For the scout's attention must be divided between his front and his rear, with a consequent loss of power of concentration. Cover which will conceal him from both directions is hard to find; and, if he be discovered, his retreat must be circuitous. His best course is to hide and wait for the departure of the party in his rear, or for night, unless he should have sufficient confidence in himself, or his horse, or his luck, to risk a chase in the open. If the discovery of his hiding-place be imminent, he should, of, course, be prepared to make a dash for escape, and should endeavour to make his appearance as

---

128. Again, more support for having well-trained officers who can develop reasoned plans.

sudden as possible, and his route one that is likely to be unexpected.

With regard to this matter of hasty retreat and the occasional necessity for it, a good scout will not willingly enter in daylight any place which is naturally or artificially enclosed, unless he is reasonably satisfied either that his retreat by the original entrance is secure, or that there is another outlet which can be used as an emergency exit. He will avoid, for example, the ground between a cliff and an unfordable river—no uncommon combination—or a continuous, steep-sided valley, or a field with fences too big to surmount and too strong to be made passable quickly. He will take care, in fact, that his plan for immediate retreat, should that become necessary, is always complete, the line of retreat being not necessarily to the rear, but in any direction which promises to lead to temporary and so to ultimate safety.

There are a few details concerning retreat that should not be neglected. If a scout leaves his horse, for instance, it should be in such a position that it can be quickly mounted, and a start can be made at a gallop. Should a bicycle be used, it should be left facing in the direction of retreat, and, if possible, in a spot (such as the top of a rise) from which advantage can be taken of a flying start. If the scout be working on foot, it is usually best to make for broken ground or woods, where a horseman would be at a disadvantage. And every scout should carry a rifle, for very often a shot or two will cause pursuers to hesitate, and thus give time to secure a start. There is no chase so keen as that after a defenceless man.

If it should be found that a good scout, either from want of confidence or from the strain caused by continual exposure to danger, is becoming inclined to linger too long at his halting-places, to observe at too great a distance, or to detect imaginary enemies, it is of no use to try to hurry him, or to correct his fault by reproof or censure. The man who is overstrained should be given a rest, which can usually be arranged by finding for him some temporary duty within the lines, or by sending him on a safe mission; the man who lacks confidence should be employed only on patrol, under an officer, so that his movements may be decided for him. By such methods of treatment, both men will probably recover; but if they continue to act independently, and yet are urged to act against what they will consider to be their judgment, although it may be only their inclination, the result is likely to be either nervous breakdown, or a reaction of angry recklessness. Scouts should be tenderly handled in this respect; it is difficult for those who have not personally been engaged in it to appreciate fully the effect on the nerves produced by continuous reconnaissance duty. The unceasing alertness required, the constant apprehension of danger to life or to success, the sensation of military inferiority, caused by working alone or in small numbers, have a wearing effect on the most resolute men; and although the determination of some may be proof against any strain, yet signs of exhaustion may be observed in the impairment of temper, of patience, or of judgment. When operating against an inactive or unskilled enemy, the strain is of course lessened; in such case it may well be found that the fascination of the game, the desire for distinction, and the pride of moral superiority to the adversary, will carry a scout through a campaign, on continuous independent duty, with no disturbance of nerve. But when the enemy is not markedly inferior, every scout should be given occasional rest or change of duty; in this way only can the standard of excellence be maintained.

The acquisition of information—particularly with reference to the kind of information which he should endeavour to obtain—although the most important part of a scout's duty and the real reason for his existence, is a matter on which it is very difficult to state any general principles. For if the scout is engaged on reconnaissance with no definite instructions, he must judge for himself in how far the information he is acquiring is likely to be of use to his employer, and to enable him so to judge, he should have some military knowledge. The better his education as a soldier, the more valuable should be his information, for general military knowledge will not only enable him to recognise unhesitatingly any important information but will point out to him the direction in which further investigations may be most profitably undertaken. For example, a scout discovers the enemy posted in a continuous line across his front. If the scout be a man of little military knowledge, he will probably return and report that at such a place and time he found the enemy across his front. But if the scout should happen to be a staff officer, the investigation would take a different form and would have a different value. A continuous line of hostile posts would to him mean outposts, and his first endeavour would be to discern or divine what these outposts covered. If the configuration of the ground favoured the idea of a defensive position, he would look for entrenchments, for signs of men at work, for guns. If confirmed in his supposition, he would consider how, if he were the enemy, he would occupy the ground; he would look for a "key" to the position, for its strong and weak points for defence, for favourable or possible lines for attack, and he would try to discover the extent of the position and to locate the enemy's flanks. It is not impossible that an acute officer might form a valuable opinion on all these points without approaching any closer to the enemy, or running any heavier risk than the man who, from lack of knowledge, had to return with nothing but the certainty that he had met the enemy. It may be admitted, of course, that the knowledge that the enemy was in a certain place at a certain time may be of great value; but if the reports of scouts are altogether confined to simple and definite statements of fact like this, it is evident that much valuable information will be lost or delayed. When once the first stage of the education of a scout—the taking care of himself—is complete, then his value is likely to increase in proportion to the extent and accuracy of his military knowledge.

When a scout is sent on a definite mission, and successfully accomplishes it, there is still no certainty, if the man has no military knowledge, that further valuable information has not been missed. He may not understand what he sees; he cannot be expected to appreciate the bearing which a seemingly unimportant matter may have on the conduct of a campaign. He may see the enemy, but may be quite unable to guess what the enemy is doing; he may note that a bridge is being repaired, but fail to connect this work with a probable desire to use the bridge; he may find stacks of supplies at an unexpected point, without thinking of the men they are destined to feed.[129] Had a civilian been sent, as Major Colquhoun Grant was sent, to find out whether the French were advancing to attack Ciudad Rodrigo,[130] he might have reported the movements of the French columns as accurately as Grant did, and this report would certainly have led Wellington to believe that the attack was imminent. But it would hardly have struck a civilian as necessary to incur grave risk in order to ascertain that Marshal Marmont[131] had left his scaling ladders behind him, although that discovery proved that the

---

129. There is a difference between training and education. Training is the ability to carry-out a task, whereas education is the ability to think and analyse. It is education that Henderson is talking about in this context.

130. Ciudad Rodrigo is a small city in the Spanish province of Salamanca that was at the time a fortified city. The Duke of Wellington laid siege to it for 10 days, capturing it from the French.

131. Auguste Frédéric Louis Viesse de Marmont (1774–1852) was a French nobleman who was also a general with the rank of Marshal of France.

Marshal had no design against a fortified place. Wellington's system of employing officers with a knowledge of war on independent and solitary reconnaissance was most probably instituted for the reasons here given, and there can be no doubt that the good officer, when he has mastered the preliminaries of his work, makes the most valuable scout. For when he succeeds, he succeeds to some purpose, and even when he fails in his main object, he will usually, in the course of his otherwise abortive effort, pick up some details, or form some opinions, which will be of use to his commander.

Apart, however, from the value of general military knowledge, there are some particular points to which the attention of scouts should be specially directed. These are principally connected with the interpretation of signs of the enemy, or with speculations as to his strength, based on limited data. Tracks, which have already been mentioned in connection with the aid they give discovering an enemy, are important in this respect also, and their study, with regard to their identification and the evidence they are of the enemy's movements, should not be neglected. A scout should know the measurements of the wheel tracks of the enemy's guns and wagons, and any peculiarity in the shoeing of his horses. He should be able to judge, by the state of the ground on which tracks are found, whether a large or a small force has passed. The track of an enemy's patrol also may inform a commander that the positions or movements of his force have been observed, and that the element of possible surprise of the enemy must therefore be eliminated from his calculations. Tracking, however, is an art the value of which is usually, for regular warfare, overestimated, and a high degree of skill in this respect, although useful, is not an absolutely indispensable part of a scout's qualifications.

The dust caused by motion is one of the most useful indications of an enemy; it can be seen at great distances, and its volume gives some idea of the magnitude of the force which causes it. The dust raised by horses moving rapidly can usually be distinguished from that which accompanies a marching column of infantry: it rises higher and is more transparent. It is not so easy to recognise the presence of guns or wagons by the dust which their movements create: it varies in appearance between the mist which floats over cavalry at speed and the cloud which chokes the tramping foot-soldiers. The chief importance of these signs is that dust often betrays the movements of troops who are otherwise concealed; usually, also, although not invariably, the direction of the movements can be discerned.

For the estimation of the numbers of a visible enemy, the time test, when it can be applied, gives, as a rule, the most satisfactory results, for the pace at which troops march varies very little. The usual march formations of the enemy should be known, as the observer cannot always be close enough to distinguish these: the calculation of the numbers which can pass a fixed point in a given time is then an easy matter. The rates of march of the different arms, and of mixed bodies, are, or ought to be, known to all soldiers; they may, however, be included here. "Combined training" gives the following: "For each minute, the following numbers would approximately go past: Cavalry at a walk-in fours, 120; cavalry at a trot in fours, 250. Artillery guns or wagons at a walk, 5. Infantry in fours, 200." Bodies of mixed troops exceeding 5,000 men, on a single road, seldom average more than 2.5 miles an hour.[132] The observer must judge for himself whether the troops are marching freely, or whether their pace has been, for some reason, checked.

---

132. i.e., about four kilometers per hour.

The strength of troops when not in column of route, must be estimated by eye. For this, the scout should have a "unit of estimation" for each arm, such as squadron, battery, battalion, and should endeavour to distinguish and count these units, rather than to hazard a guess at total numbers. Skill comes by practice. The scout should know the appearance of his selected "unit of estimation" at all distances, and in all formations. If he can reckon up, with some accuracy, the strength of a force of his own army, the scout will have little difficulty in adapting his knowledge to the slightly altered conditions required for the observation of any other civilised army.

The method of counting bivouac fires in order to calculate the enemy's strength cannot be recommended. There are few artifices more common in war than the lighting of dummy fires.[133] Nor is there any real basis for calculation. The same fire may serve ten men on one night, and a hundred on the next, and there may be as great a variation between two adjacent fires. If a commander desires to magnify his forces, the number of his fires need be limited only by his fuel; if he desires to minimise his strength, he can decree that half his men shall eat cold rations. It may be sometimes gathered, therefore, that if the enemy's fires are conspicuous, it is likely that the hostile commander wishes to exaggerate his force; but beyond this deduction, there is not much of value as regards numbers, to be extracted from observation of the cooking arrangements of an adversary. The

---

133. This is a common counterintelligence technique known as *deception*. Commanders throughout the ages have used this ruse. See Hank Prunckun, *Counterintelligence Theory and Practice, Second Edition* (Lanham, MD: Rowman & Littlefield, 2019). Perhaps the first recorded use of campfires to exaggerate troop numbers was used by the Byzantine general Belisarius during the Gothic War (535–554). See Basil Henry Liddell Hart, *Strategy, the Indirect Approach* (London: Faber & Faber, 1954).

"watch-fires of a hundred circling camps" may have a message for the poet, but they convey very little to the scout. The appearance of fire or smoke, however, may indicate the presence of the enemy, and this alone is sometimes useful information, especially as a guide to further investigation.

There is one detail of information about the enemy on which scouts should always be able to report with accuracy—that is, the efficiency of his system of protection. It is with the hostile outpost line or covering troops that the scout usually comes in contact. His success depends very much on the penetrability of the barrier, which is designed by the enemy to ward off intrusion and prevent surprise. Knowledge of the enemy's protective measures is, therefore, almost forced on the scout. He can hardly fail to discover whether the enemy's outposts are vigilant or are careless: whether the enemy's patrols are bold and alert or are timorous and unobservant. And this knowledge may be of the first importance: great issues often depend on the efficiency of outpost service. A slackness in the enemy's outpost line which permits a scout to achieve a successful reconnaissance, may also enable the scout's commander to achieve a victory by surprise; and the report as to the neglect which enabled the scout to succeed may be of more value than all the rest of the information which his success enabled him to gather. Therefore, the report of any reconnaissance which comes into contact with the enemy should include information as to the efficiency of his protective screen.

The information which enabled a commander to deliver a successful night attack on one of the British columns in South Africa, was obtained by two Boer scouts who had approached the British bivouac in the hope of being able to remove horses to remount themselves.

So far, only direct information—that is, information obtained by the evidence of the scout's own senses, or of his reason—has been touched upon. Indirect information—that obtained second-hand from others—opens a much wider field: so wide, in fact, that here the object can be treated only in its most practical aspect. For the obtaining and sifting of indirect information belongs properly to general intelligence work and is only incidental to reconnaissance. The indirect information obtained through the medium of scouts is but a small fraction of that which reaches a properly constituted intelligence department from all sources.[134] Nevertheless, the scout cannot afford to neglect the hearsay information he may glean when on reconnaissance. It may be of importance to the scout and assist him in his effort to obtain direct information, as when it touches on a possible route to be followed, or points to a hidden danger, or it may be connected with other matters which are of importance not to the scout but to the army.

There are four sources from which a scout, when actually on reconnaissance, may hope to obtain indirect information: these are, the inhabitants of the country, prisoners who may be captured, deserters from the enemy who may be met with, and documents which may come into his possession.[135] With regard to the inhabitants, the possibility of extracting useful information from them depends almost entirely on their attitude towards the combatants. Much help may be obtained from individuals of a friendly population, and on occasion from neutrals. The scout will find, however, that he has, as a rule, but little time to spare

---

134. Known as *all-source intelligence*. That is, intelligence gathered from human sources, aerial and satellite imagery, electronic and electromagnetic measurements, radio signals, and open-source data.

135. And, I would argue *artifacts* should be added to this list. For instance, weapons, uniforms, and other enemy equipment of all kinds.

for elaborate questioning on matters other than those closely connected with his immediate mission. Detailed examination of inhabitants, and also of deserters and prisoners, on general subjects, can be carried out better in camp and at leisure: the scout may be satisfied if he obtains something to help him on his way. He should, when possible, take measures to enable the staff of his army to get into communication with any person whose information he thinks may be valuable; and if there is no possibility of this, he should gather as much as he can; but his efforts will naturally be directed first to his own affairs. The type of helpful information which he may expect from friendly inhabitants is, first of all, topographic; this is often of great assistance. He may get a fair notion of the "lie of the land" in unknown country; he may secure a guide who knows the bypaths of the neighbourhood; he should at least be able to verify his position and some useful landmarks. If hostile forces be near, he is likely to glean something of their position, and perhaps of their strength, although in this latter particular, civilian information cannot usually be trusted. And if the enemy's patrols should be active in the neighbourhood, he may learn something of their customary mode of procedure. From neutral inhabitants he might get similar information, but with more difficulty. From a hostile population the scout will get neither information nor aid, save only that assurances of safety, from a hostile source, may serve to warn him of imminent danger. His safest course is to consider hostile inhabitants as part of the enemy, and to avoid them altogether.

Scouts working alone or in pairs are not very likely, and are seldom willing, to capture a prisoner, as in such case the reconnaissance must be interrupted until the captive can be handed over to safe keeping. Nor should they have any communication with apparent deserters, if they can avoid them; but such men should be observed, as far as possible, and what is seen of them should be reported. For spies often assume the guise of deserters; and the more that is known about them, and especially about the method by which they left the enemy's lines, the better.

The information contained in documents which may come into the possession of a scout on reconnaissance, is likely to be of more use to the commander of the army than to the scout. Any document which appears to be of possible value should be handed over with the scout's report, on his return. A single instance will be sufficient to illustrate the possible value of stray documents. On September 13th, 1862, "fortune put in M'Clellan's [sic] way such an opportunity as has rarely been vouchsafed to any general. A Federal private discovered wrapped around a handful of cigars a copy of Lee's orders to D.H. Hill, giving full particulars of the intended movement against Harper's Ferry, and detailing the positions which the different portions of the Confederate Army were to occupy for the next four days. Thus, in an instant it was revealed to M'Clellan [sic] that his foe had divided his army, and that it was in his power to concentrate against either half an absolutely overwhelming force."![H, 136]

So much for the independent scout. The action of patrols and larger bodies on reconnaissance will be discussed separately, but it should be remembered that the more nearly the individuals composing any reconnoitring force approach the standard of the completely equipped scout, the better for the success of the enterprise.

---

136. Walter Birkbeck Wood and James Edward Edmonds, *Civil War in the United States, 1861–65* (New York: Putnam, 1905).

CHAPTER VI

THE PATROL

Patrols are used in war for many different purposes and are organised in as many different ways. The only common factors which apply to all patrols are that they are detachments of small size compared to the total number of troops on either side, and that they carry out their duties while moving from place to place. These duties may be for preservation of order, as in a camp or town; for the protection of some line too extended to guard with fixed posts, such as a railway or road; for maintenance of communication between separate bodies of troops; for information; and no doubt for other special purposes. The most frequent, as well as the most important of these duties, is that of the pursuit of information; and for this purpose, there are three classes of patrols, as there are three methods of reconnaissance—protective, contact, and independent.

It is to the last, the independent patrols, that attention must first be directed; for if the art of leading them successfully be acquired, the command of any other type of reconnaissance patrol will be easy.

The duties and difficulties of an independent patrol are not quite the same as those of an independent scout, and when any reconnaissance enterprise is to be undertaken, the conditions of the case should be closely examined, in order to determine whether the employment of a couple of scouts, or of a patrol, or perhaps of a brigade of cavalry, will promise the best results. The conditions which are of general application in the choice between scouts and patrols are but few and may be briefly outlined as a guide. The scout, or pair of scouts, would as a rule be preferred when concealment is essential, and usually when concealment, although not essential, is possible, throughout the reconnaissance. A patrol would be used if information is to be sent in at intervals, or when it is desired to capture prisoners; it should usually be chosen if concealment is impossible, or if the expedition may last so long that the scouts will require relief. Patrols also can work more expeditiously than single scouts and are therefore preferable for urgent missions. These are not rules, but general guides; and it will be found in practice that there are special conditions which will also influence the decision as to the method to be used. The individual preference of the most efficient scouts ought to be considered, if possible. Some men lose confidence if attached to a patrol; some dislike the responsibility of uncontrolled action. The methods of the enemy must be considered: should he habitually reconnoitre with strong patrols; it may be easy for single scouts to evade these. In many cases it may be advisable to use a patrol for the first part of a reconnaissance, and at a certain stage to break it up into scouting parties. When it is not expected that the enemy will be met with, or in a reconnaissance of a simple nature, when, for instance, it is necessary merely to make certain of the absence of the enemy from a particular locality, it is better, as a rule, to use patrols, for patrol duty is less arduous than singlehanded reconnaissance. But to this also there are exceptions, as when there are many scouts and few horses. Conditions such as these should be considered for each case.

The ideal reconnaissance patrol is that which is composed of trained individual scouts and commanded by an officer who is himself qualified as a scout. Such a party, however, is not always available. It may be well, therefore, to consider the proper organisation of a patrol, and the duties of those who take part in it, that there may be some guide as to the principle on which men should be selected and duties apportioned. There is first the commander. He should have rank in order that he may have authority; for his decision must be final as to every detail of combined procedure. There are cases also in which it is necessary to impart to the leader of a patrol information of a secret nature, perhaps connected with the plans of the general commanding—information which cannot properly be entrusted to any but an officer of proved discretion.[137] If no officer or non-commissioned officer be available, then the leader should be selected for such qualifications as are for the particular service most important.

Next in importance is the observer. His selection depends, to some extent, on the nature of the information required. He may be a naked savage, taken out to identify a track; he may be a topographist,[138] required to report on ground, or an artist, to make a freehand sketch or panorama; or he may be the commander-in-chief himself, bent on personal reconnaissance.[139] These, of course, are extreme cases. The commander will usually be also the principal observer.

Next in order are the flankers, and it is the presence of these that differentiates the patrol from the scouting party, for the addition of flankers necessitates central control. These must be at least alert and intelligent men, who can understand instructions and obey orders. It is their duty to look out to the front and flanks, and to prevent the interception of the observer.

---

137. If captured, this information can be used to deduce the force's plans and perhaps, its order of battle.

138. Now referred to as a *topographer*.

139. Compare to the photographic aid that is now available where images can be sent to the command as events occur from ground or airborne scouts.

To these may be added, as required, an interpreter, orderlies, and escort.

From this analysis it will be seen that the minimum strength of any independent patrol is three men—an observer and two flankers—and where all three are trained scouts, such a patrol may be effective enough. There is an evident advantage in combining the duties of observation and command. Any or all of the three scouts may be efficient interpreters; and orderlies or escort are not always required. The chief difficulty which is experienced in conducting a patrol of three is that of communication. Smooth working is only possible if all three are experienced men, accustomed to work together and to follow similar methods. The addition of a fourth man, interchangeable with either of the flankers, removes this difficulty, and in other ways strengthens the combination; the group of four is, indeed, considered by many good authorities to be the ideal patrol for ordinary circumstances. Theoretically, it is so nearly so, that it may here be taken as a model for the demonstration of the proper working of a patrol under varying conditions.

It may be said at once that the principles which should guide the operations of a single scout are often applicable, not only to the conduct of the separate individuals of a patrol, but to that of the group working in combination. Thus, it may happen, particularly when concealment is possible, as at night, that a patrol leader will find that his best chance of success is in closing in his men and in following, as far as possible, the methods he would pursue if alone. At such times, in fact, his companions are an encumbrance; they add to the risk of discovery and do not directly assist progress. The disadvantage of numbers may, however, be lessened by the observance of a few simple maxims which, although almost axiomatic, are sometimes forgotten or neglected.

If a patrol be occupying a post of observation, at or near which cover can be found, the number of observers allowed to look out should be limited, the remainder of the party being, if possible, completely concealed. In many places a single lookout man will be sufficient, both for information and for security. A man must expose himself, to some extent, to obtain a good view; the more men, therefore, who are allowed to look out, the greater the risk of detection. There are occasions when the necessary field of observation cannot be commanded from any single point; then lookout men must be posted wherever necessary, but on manoeuvres, or even in war, it is not unusual to see the whole of a patrol occupying observing points from all of which only the same ground is visible, each man craning his neck to see more than his neighbour. It is the natural tendency of every man on dangerous service to keep a lookout for himself; but it is the duty of the patrol leader to make the necessary dispositions for observation and for security, and to ensure that neither from nervousness nor from curiosity are his arrangements unnecessarily duplicated. If the observation posts are skilfully chosen, it will usually be possible to conceal the unemployed men in such a manner that, while they can see or hear their own observer, they are hidden from the view of any possible enemy.

When cover is only partial but is yet so good that the patrol leader considers his best chance of success lies in the probability of concealment, the formation to be adopted, at those times when the party cannot be completely hidden, requires consideration. If it should be necessary, for instance, to cross an open space between two woods, under the possible observation of a distant enemy, there are three methods which at once suggest themselves; to cross, man by man, in succession; to move in a body; to move simultaneously but dispersed. Of these, the first method cannot be recommended. Of all forms of movement, that which is most likely to attract the eye is recurring movement. A single movement may be seen but may be at an end before the mind of the observer has fully grasped its significance; he may be left in doubt as to whether he really saw anything or not. A repetition of the movement brings certainty. The difficult passage should therefore be affected by the whole party simultaneously. Whether they should move in a body or dispersed depends a good deal on the light and on the background, and also on the presence or absence of horses. If the enemy be not very near, closed bodies, even with led horses, if moving slowly, will often escape observation, or, if observed, will be difficult to identify. In a failing light, men widely extended may well pass unnoticed. If the distance be very short and the light good, a quick rush may offer the best prospect.

Should the patrol be so near to the enemy that its movements may be detected by either sight or hearing, as might happen at night when there is a moon, or when the fixed beam of a searchlight must be crossed, the choice is not so easy. Silence and accuracy in night movements are best secured by adopting the formation of single (Indian) file; yet one feature of this formation is repetition of movement. In such case, the leader must judge, from considerations of ground, weather, and the position of the enemy, whether he is more likely to be seen or to be heard and must adapt his procedure accordingly. Especially in an imperfect light is it desirable to avoid repetition of movement; prolonged scrutiny is required for the detection and identification of objects in semi-darkness, and the repeated presentation of the same performance is then, more than ever, a valuable aid to a hostile observer. Single file, in such case, is not a suitable formation, except for a direct advance towards the enemy, and then only if the men are well closed-up. If crossing the enemy's front, a patrol should move in a group, or in line, shoulder to shoulder. Any man by man movement in either direction is dangerous.

The close order formations here advocated for night work are suitable not only for concealment but for safety; for in darkness or semi-darkness, a group offers no better target than does an individual; accurate fire, moreover, is seldom to be expected. And in any sudden encounter at close quarters, unless against overwhelming force, there is a tactical advantage in concentration. The only occasion when the dispersion of a night patrol would seem to be judicious (except when at the halt, to keep the necessary watch), is when retiring after attaining the object of a reconnaissance, so that, in case of interception by the enemy, there may be a probability of somebody escaping with the information which has been gained.

A patrol at night must conform instantly to the movements or signs of the leading man, whether he be the commander or not. The leader may be chosen because he can act as guide, for his skill in selecting practicable paths, for his acute hearing, or for any other reason; but if he is not the captain, he is at least the pilot, and must, for the moment, be obeyed instinctively. If the commander of a patrol objects to delegate his authority in this way, he should himself lead. For the leader is in a better position to see, or hear, or feel, than the others. He should be the first to suspect or discover danger. He can pass signals easily by touch; and if he halts, the party naturally closes up. The front is, in fact, the position for command while on the move by night. It is sometimes possible for the commander of the patrol to be beside the leader; but this arrangement invites hesitation, and even discussion, both of which are to be avoided.

If it be necessary at any time to detach a man from a night patrol to reconnoitre to the front or flanks, the patrol should always remain halted until his return, unless it should be quite immaterial whether he re-joins or not. For if the patrol follows the scout without waiting for his report, the whole party might just as well have moved in the first instance, while if it be

122

intended that the scout and the patrol should move by separate routes to a new rendezvous, the difficulties of the reconnaissance are being wilfully multiplied. It is hard enough to find one suitable path at night, and much more so to find two which will meet at a certain point; moreover, when a party is separated, an enemy may be mistaken for a friend. For tactical purposes, such as the surrounding of a house, separation is, of course, permissible; and, in such cases, there is usually a point of concentration which can hardly be missed. But, generally, the safety and success of a night patrol are best ensured by its remaining as much as possible concentrated: that is, by its imitating the methods and procedure of a single independent scout.

For reconnaissance in daylight and in open ground, the advantages of the patrol, as compared to the single scout, are apparent. Scouts advancing unconcealed towards an active enemy may be detected at a distance and must expect that efforts will be made to ambush or intercept them. To evade these attentions, dispersion is necessary, in order that the flanks may be watched, that some mutual support may be possible, and that complete interception may be made difficult. But in order to retain unity of purpose and of command in a patrol which moves in extended formation, the principle on which each man is to work must be understood, and for this reason, some details of methods which have been practically tried may be useful.

First of all, it may be said that no formation is entirely satisfactory which does not provide for verbal communication between the commander and the flankers. Signals are not always sufficient. For the simplest communications such as "safe to proceed," or "imminent danger," or to summon the commander, signals are convenient, and may be necessary; but no signal can convey the details of information, and it is often in the details that the importance lies. And if commander and scouts have themselves to keep moving back and forward in order to communicate, there is a loss of time and a lengthening of the journey. The following is a simple method of communication, suitable for the purposes of any patrol of four or more men: it may be called a good general arrangement.

The patrol consists of an officer, who is the principal observer, and three trained scouts. In the normal formation, the officer moves in the centre, accompanied by one man, and a scout is thrown out on each flank. Certain signals are necessary; one for the commander to close the patrol on himself, another for a flanker to notify that he wishes to come to the commander to report, a third for a flanker to call the commander to join him in order to observe. Should the commander desire to communicate with a flanker without closing, he has the spare man as messenger. The first signal explains itself; it may be wanted at any moment. If the second signal be made, the spare man at once leaves the commander, and moves off to relieve the flanker who made the signal. The flanker may wait to be relieved, or may meet his relief half-way, according to circumstances. In the latter case he would, of course, pass on to the relieving man such information as might be necessary. He then joins the commander and becomes spare man, communicating his information verbally, and remaining in the centre until a flanker again wishes to communicate. If the third signal be made, the commander moves to the required flank and observes for himself, the spare man remaining in the centre as pivot of the patrol.[140]

This method gives, for all practical purposes, complete communication within an extended patrol. It also has the effect

140. Although scouts could use two-way radios with a laryngophone and covert earpiece, this arrangement is suitable where radio transmissions are subject to enemy intercept.

of saving time, as no halt is necessary, unless imposed by caution, or by a desire for prolonged observation. These advantages are due to the presence of the spare man. The commander of a patrol, therefore, should always endeavour, whatever formation he may choose to adopt, to retain at least one man as the necessary link in his system of communication. From this it appears that the theoretical minimum strength of three men is insufficient, and that to be effective a patrol should consist of at least four men, and the formation suggested above—two in centre, one on each flank—is probably that which in open ground best combines the requirements of observation, security, and control. Any deployment which separates all four men, makes it impossible to hold any communication, except by signal, without breaking the formation.

The "diamond" patrol, formerly advocated in some textbooks, fails in this respect, and is not suitable for parties of less than five, or, if the commander is to move in the centre, six men. This formation, indeed, shows the danger of too much theory in matters pertaining to war; it is simply the application to a patrol of the system of protection commonly adopted for bodies of fighting troops.[141] For reconnaissance work it is pedantic and unpractical. Experience.

In considering this question of formations, there are some principles which may serve to distinguish between those which are suitable and those which are not. For one thing, any system which provides for the permanent protection of the commander is bad. The commander of a patrol should take the same chances as his scouts, at any rate until the object of the reconnaissance has been attained. He may not be so skilful as his assistants, but he

---

141. Or, stated another way, experience can be a person's best teacher; so, do not ignore the lessons learned from what went before.

must be as intrepid, and show it, if he expects to keep the standard of efficiency unimpaired. It is no uncommon event for scouts to conspire for the protection of their commander, especially if he be new to the work; and however admirable such spirit may be in itself, the acceptance of protection is bound to result in loss of authority. The place of the commander is therefore in the front line when advancing, in rear when retreating; if there be any special risk to be taken, the post of danger is his by right, and he must claim it. This does not mean that the commander should always lead. The manoeuvres of a patrol are influenced chiefly by ground, and in practice, one or both of the flankers will frequently be in advance of the centre. The point is that it is both unnecessary and inadvisable to have an advanced scout in front of the commander.

Another principle is that the formation should be elastic: that is, that the various units of the patrol should be able temporarily to diverge, or to alter speed, without throwing the formation out of gear. A scout cannot carry out his duties properly if he is obliged to conform closely to another's movements. He must conform generally, of course, to the movements of the commander, and this is easy enough when the formation is quite simple, and when the commander is in front line. In fact, elasticity is attained when the commander and the responsible scouts move normally in line, for thus only is it possible for each to observe without difficulty, and even sometimes to anticipate, the movements or signals of the others. If a scout be placed permanently in advance, unless he be the commander, he must look back for signals or guidance, and this not only distracts his attention, but necessarily, unless the commander is content to be a cipher, limits his discretion in movement. The "diamond" formation, for example, with the commander in the centre, can give no satisfaction unless the whole responsibility of the patrol be thrown on the leading scout, for otherwise he will be so occupied in looking for guidance from the rear that he will find it impossible to do a scout's work, and will become merely a passive shield to the commander, an advanced target for a possible enemy. Even if he be given a free hand, he must watch both front and flanks, for the flankers, being to the rear, can give him but little assistance. There is no elasticity in such an arrangement.

These two principles, if admitted, are additional reasons for the selection of the line formation as the normal disposition of a patrol in open ground, and attention may now be directed to some of the circumstances which would necessitate temporary modifications of this formation. These are principally connected with the approach to or searching of places where enemies might be concealed, awaiting an opportunity to attack the patrol. Such are defiles, detached woods, villages, or the reverse, slopes of hills; and although the details of procedure in each case must be governed by the particular circumstances, it will be found in practice that there are certain methods of dealing with such difficulties which are applicable to most conditions. For instance, if the dangerous spot is not very great in extent, and yet is an obstacle either to movement or to sight, then it should, if possible, be first reconnoitred from a flank, or at least by a flanker. That is to say, that a patrol should always endeavour to skirt rather than to penetrate an obstacle which might divide it so that free communication would be even temporarily lost. The accepted method of reconnoitring a village or defile by sending flankers round both sides and passing an advanced scout through the middle, is suitable only for contact reconnaissance, where support is at hand. For independent reconnaissance, this system is unsuitable, because it exposes too many men to simultaneous danger, in addition to the fact that there is usually a difficulty in

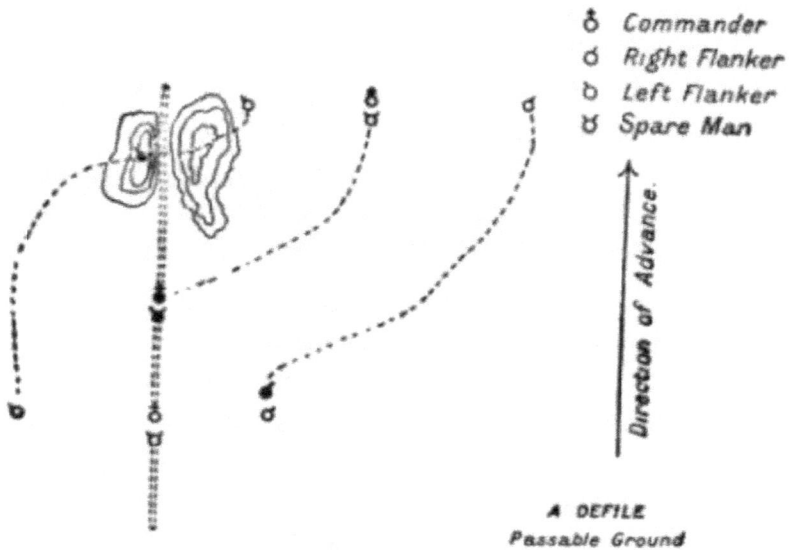

δ̂ Commander
δ Right Flanker
♭ Left Flanker
♉ Spare Man

Direction of Advance.

A DEFILE
Passable Ground

retaining control. It is preferable, in conducting an independent patrol, to expose each man alternately to danger, securing the safety of the remainder, rather than to endeavour to lessen each risk by sharing it among all and losing the advantage of combined action. The difference between the procedure of contact and of independent patrols may be made clear by examples.

It may be necessary for a patrol to ascertain that a certain defile, such as a steep-sided valley, through which runs a road, is clear of the enemy. A contact patrol would probably act in much the same manner as an advanced guard, which, in miniature, it really is; it would crown the heights on both sides and pass a scout or party along the road. The method suggested for an

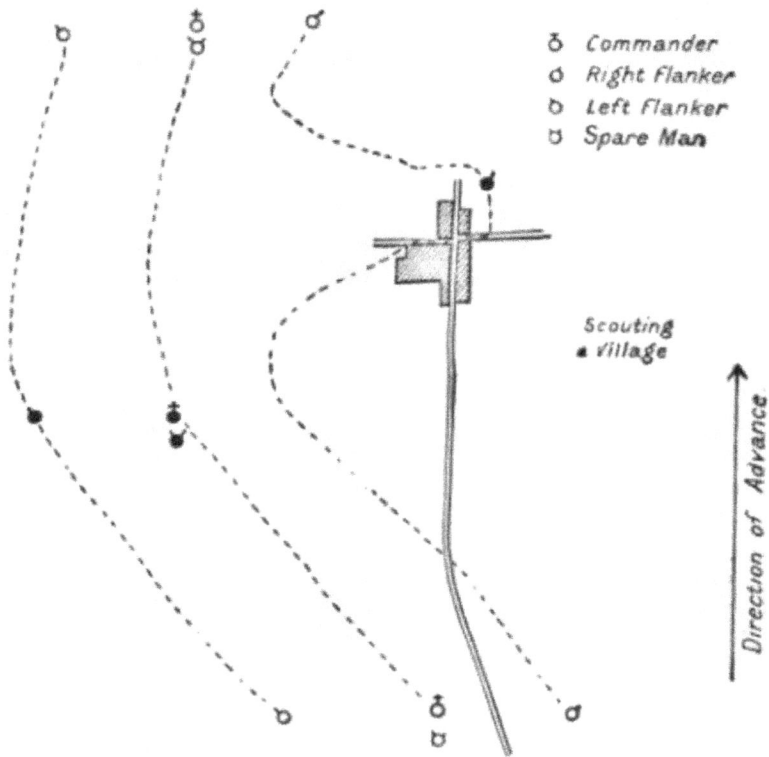

δ Commander
ο Right Flanker
ϸ Left Flanker
ʊ Spare Man

Scouting
▲ Village

Direction of Advance

independent patrol is to wheel to right or left and leave the searching of the whole defile to the flanker who is brought nearest to it; and he would move across, or round, or through it as he thought best. When there is more than one possible line of retreat, it may be unnecessary even to wheel; a simple divergence, which will bring one flanker in front of the spot to be searched, may be sufficient.

If there should be good reason to expect the enemy in a particular spot, a different system may be required; in such case, the commander of the patrol would probably share the dangerous duty; but for the ordinary procedure of a long reconnaissance, this method is speedy and convenient, and with a small party is probably also the safest for all concerned. For not only can the safety of the majority be guarded, but also, so long as the power of combined action is retained, there is the possibility of taking steps to protect, or at least to cover the retreat of, the single man exposed to danger. The convenience and certainty of manoeuvre due to the possibility of retaining the same or a similar formation under varying circumstances, is important. When the principle is thoroughly understood by the men of a patrol, no special instructions are necessary; each duty as it arises is undertaken by the assigned man without delay or discussion.

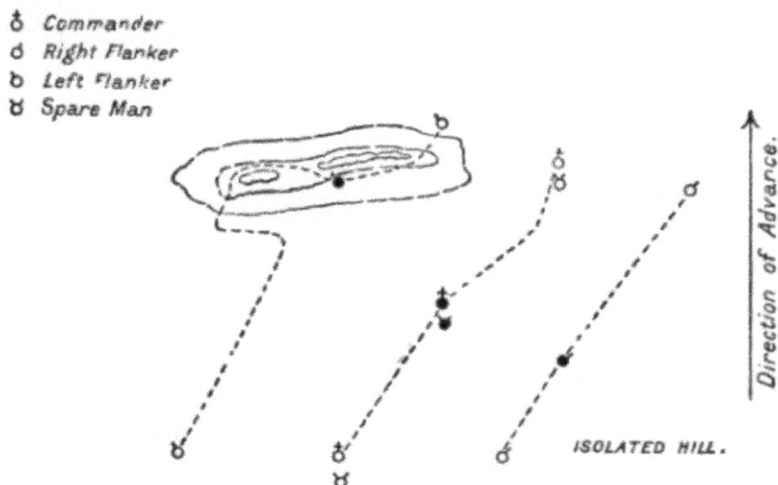

There is also an advantage in making a practice of scouting from a flank, for if the patrol has been observed by the enemy, and an ambush prepared, the enemy's arrangements will probably have been made with a view to meeting a direct advance, and any attempt on his part at rearrangement is likely enough to lead to his detection.

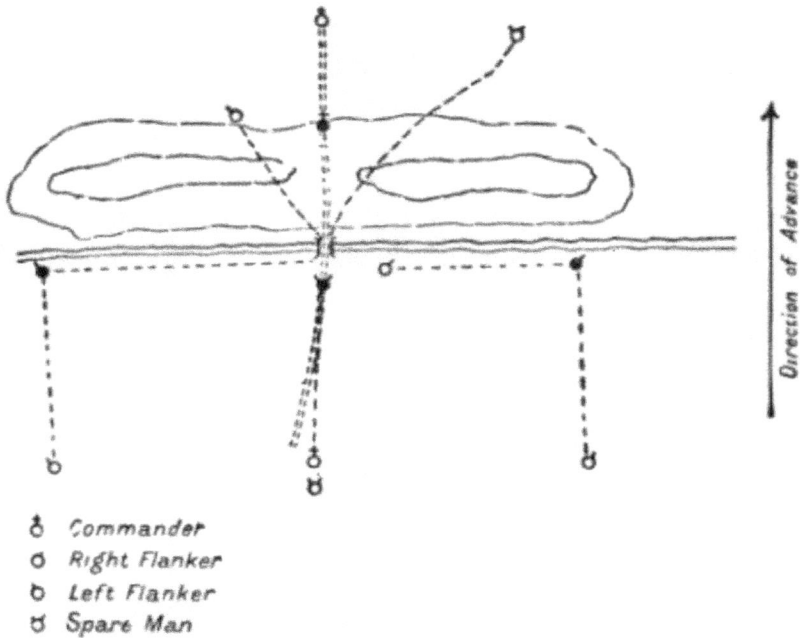

ᕣ *Commander*
ᕯ *Right Flanker*
ᕲ *Left Flanker*
ᕳ *Spare Man*

Bridge.

Commander scouts. Spare man becomes right flanker.

A continuous feature which might hide the enemy, such as a long ridge, or a river, should, if it runs obliquely to the line of advance, be examined by the flanker who first approaches it. Should it be perpendicular to the line of advance, the commander must decide whether he will wheel, so as to bring it within the sphere of one or another flanker, or whether he himself or his spare man shall advance to scout it.

Should the feature be also an obstacle, with only a single passage, it may frequently be convenient to make the first reconnaissance from the centre, the flankers delaying until the front is reported clear (see figure, "Bridge" defile). The daylight passage of a large wood by a patrol requires great caution. For, in most woods, if the patrol be extended, control and communication are difficult; if the patrol be closed, the advantage of flank observation is lost, and should the patrol be discovered or ambushed, every man comes simultaneously into danger. The commander must handle his patrol according to the conditions of the moment: whenever the wood is fairly open, he should extend; when the density increases, he should close. As a general rule, the advantages of extension should be sought, and the front should be as great as is compatible with due control. The circumstances may, of course, be such that concealment is the main object, as in the case of a patrol which is endeavouring to pass unobserved close to a known post of the enemy. The action of the patrol would then approximate to that of a single scout; for such enterprises, being dependent for success on concealment, should properly be entrusted to scouting parties, and not to patrols.

When no extension is possible, as in the case of a patrol following a path through otherwise impassable jungle, the formation to be adopted depends chiefly on tactical considerations. There should be at least two of the party in front, as it is likely enough that if the enemy be met with under such conditions, only one or two men on each side will be able to come into action at a time, and the combat, at first, may not be unequal. Another point to be remembered is that a man should be kept so far in rear that he is unlikely to be involved in the first affray and has therefore a fair chance of making his escape with any information that may have been obtained.

The principles which have been suggested for the conduct of a model patrol of four persons are equally applicable to parties of any strength up to a dozen men. Patrols of greater strength than this can hardly be required for purposes of reconnaissance alone; if strength be required for security, or to overcome opposition, then the men added for this purpose should be looked on as escort, and not as an integral part of the patrol.[142]

When it is necessary, because of the hostility of inhabitants, or for other reasons, to employ independent reconnaissance parties of a strength greater than usual, the principles on which they should be conducted are not altered, although it may be necessary or advisable to adopt methods more crude and decisive than would be suitable for small patrols. Concealment is difficult not only on account of the size of the party, but also because hostile inhabitants will certainly, if they can, give information to their own troops. Success in evasion, therefore, depends chiefly on swiftness of movement, and swiftness is incompatible with careful scouting of the ground to be traversed. Traps and ambushes[143] must, to some extent, be risked; and the prospect of having to fight for safety will be ever present in the mind of the commander. On the other hand, the strength of the party will enable it to deal with the enemy's ordinary patrols, and any concentration against it should be detected, so that the risk of ambush or interception is not so great as might appear. The real danger is that the party may stumble on a superior force of the enemy already concentrated and find difficulty in extricating itself. The consideration of the best course to pursue under these circumstances is perhaps a matter rather of tactics than of reconnaissance; but it may not be out of place to suggest here that a bold front, a wide extension, and a lavish expenditure of ammunition, are useful preliminaries to a rapid and concerted retreat. And with regard to retreat, the same general rule applies

---

142. This is more than a technical point, it is an import concept to note for planning, logistics, and operational efficiency.

143. An excellent reference on this topic is the 1965 reprint of the Australian Military Forces (since 1976, the Australian Defence Forces), *Ambush and Counter Ambush* (Boulder, CO: Paladin Press, 1981).

to all independent reconnaissance, whatever the size of the party; the line of retirement should not be the same as that by which the advance was made.

Patrols in very enclosed country frequently find great difficulty in keeping a proper lookout to the flanks. Horsemen or cyclists are then practically confined to the roads, and it is important that all roads leading to the flanks should be utilised for observation purposes. Independent reconnaissance in enclosed country is always, if effective, a slow process. If, indeed, a country is closely fenced, it may sometimes save time to carry out reconnaissance on foot. For a mounted patrol must halt at each cross-road until the lateral road or roads have been reconnoitred to such distance as the commander may think necessary, whereas a flanker on foot can make his way across country retaining his proper position in the patrol.

When, however, a long reconnaissance must be undertaken in enclosed country, cyclist patrols have great advantages, provided good roads exist, and it may be useful to outline a convenient mode of procedure for these. It is evident, of course, that such patrols can hardly hope to elude the enemy's patrols by passing between them, for in enclosed country all roads between the opposing forces would naturally be watched by both sides; but it is not impossible for a cyclist patrol to circle widely round the enemy's flank, and reconnoitre effectively on his flank or rear. The method of concealment adopted by such a patrol is that of keeping generally at a distance during the outflanking movement, and of scouting carefully up to the enemy's lines at certain selected points. These periodic investigations would be carried out probably on foot, in accordance with the usual principles; but in order to preserve the security of the party at other times, a definite system of procedure is necessary. The dangers to be guarded against are chiefly two—the possibility of meeting or being intercepted by hostile patrols, and the possibility that the enemy's flank may be extended farther than was expected, and that the patrol may strike it. To guard against these dangers, a lookout to the front and to both flanks is necessary; but the nature of the dangers is such that the intersected country between the roads is not likely to be of importance. No body of the enemy, whether moving or stationary, will avoid the roads altogether, and reconnaissance by road should therefore give fair warning of the enemy's presence.

It is desirable that a patrol which is to undertake a reconnaissance of this kind should consist of not less than a commander, an assistant, and eight scouts, in order that the cyclists may work in pairs, thus allowing one man to dismount and work on foot, leaving his cycle with his comrade. The suggested distribution is: the assistant commander and three scouts for advanced guard, and the commander and five scouts for observation to the flanks. The procedure is as follows: The advanced guard moves along the selected road, keeping a lookout to the front, for which it is responsible, and, of course incidentally, to the flanks, for its own security. The remainder of the patrol follows in a body. At the first cross-road, the commander of the advanced guard drops his third man; with the other two he continues his original route for such distance as may have been previously agreed upon, and then halts. The commander of the patrol on reaching the cross-road, decides whether or not it is necessary to reconnoitre it. If his decision be negative, he so informs the advanced guard, by means of the man who has been left behind, and the reconnaissance proceeds as before. If it should be necessary to scout the transverse road or roads, the advanced guard is informed in the same way, and the roads are reconnoitred by the allotted flankers for such distance or time as the commander directs, the extra man remaining with the commander as messenger. If no enemy is found, the flankers return, the messenger is sent to report all clear to the advanced guard, and the patrol continues its journey in its original formation, repeating the precautions at each successive cross-road.

There are evidently many possible modifications of the system here outlined. Like other systems, it must be adapted to the conditions of each case. The strength of the patrol, the proximity of the enemy, and the distance to be traversed, are all factors which should weigh with the leader in his choice of procedure.

The tactical handling of a patrol in an encounter with the enemy is not a very complicated matter. In a patrol combat where numbers are fairly equal, that party which first discovers the other has a certain advantage. By day, a patrol which is extended will probably get the better of one which is closed, as it is possible for the flankers who are extended to act against or threaten the flanks of their adversary. Patrol commanders who are forced to fight should rely on the rifle; it is usually a disadvantage to fight mounted. If the strength of any patrol exceeds six men, the flankers should be double, and the remaining men should normally accompany the commander in the centre.

It may be necessary also to take with the patrol men such as messengers, guides, or interpreters who are not fitted to take any part in the actual reconnaissance; these should accompany the commander, provided their presence does not unduly increase the strength of the patrol. If these assistants are very numerous, some of them should be dealt with in the manner provided for escorts.

An escort may be required to give a patrol a start by driving off individuals or small bodies of the enemy forming a continuous line; to protect the possible retirement of the patrol, if pursued; to provide messengers; to capture prisoners; or to deal

☽   ☿   ♂

ŏ ŏ

Patrol of Five.

♭ ♭   ☿   ♂ ♂

ŏ

Patrol of Six

♭ ♭   ☿   ♂ ♂

ŏ   ŏ

ŏ

Patrol of Eight.

☿ *Commander.*
♂ *Right Flanker.*
♭ *Left Flanker.*
ŏ *Spare Men.*

with hostile inhabitants. The last of these duties is the only one in which it is usually essential that the escort should actually accompany and form part of the patrol; under such circumstances the action of patrols is correspondingly hampered, and the results of reconnaissance unsatisfactory. In other cases, it is often possible to detach the patrol from the escort, and to move the latter separately to selected positions as the reconnaissance progresses. The escort is thus used as a kind of advanced base, through which information is transmitted, and on which the patrol commander can rely for assistance or for refuge. If the escort be a strong one, it is sometimes possible with advantage to leave communication posts, by which information may be quickly passed back to the main body; but this system is useful only if the movements of the reconnaissance can be anticipated with some accuracy, for otherwise a patrol may at any time find itself as near home as are the communication posts which it has dropped previously. These posts are advantageous in the case of a long reconnaissance which starts at some distance from the enemy; in other cases, owing to the erratic movements of most independent patrols, and the necessity of recalling the posts on the conclusion of the reconnaissance, the system is seldom suitable. The best results in communicating information are usually obtained by detailing men of the patrol, or of the escort, as messengers. These men should observe the route and the country, so as to be prepared to make their way back to headquarters when required.

The capture of prisoners is a duty frequently allotted to patrols; and even on this very practical matter, a hint may be of benefit. It is easier to catch a man by waiting for him than by chasing him. It is also safer.[144]

In the principles which should guide the procedure of protective and of contact patrols, there is much that is similar, partly because there is one service that is common to both—security. Whether it be a contact patrol feeling for the enemy, or a protective patrol verifying the enemy's absence, each is responsible for giving warning of the presence of any enemy in its particular sphere of observation. Such patrols are in no way independent; they are limited in direction, in range, in initiative. They are closely tied to the force from which they are detached; they are the sensitive ends of the feelers with which the commander of that force, whether offensively or defensively, gropes to find his enemy. Their places and their duties are assigned to them as part of an organised scheme; if one of them

---

144. This advice also applies to the reconnaissance patrols conducted by friendly forces. That is, if we accept Henderson's proposition that one is more likely to be captured in a trap if troops "run," then it follows that if a patrol suspects an ambush ahead, it has a greater chance of escape if it attempts evasion.

should lose touch with its own force, or should cut loose and act independently, then a blank is left, a feeler is wanting, the organism is maimed. The connection should be elastic, no doubt; rigidity is always undesirable on reconnaissance, but connection should never be lost, unless it is severed by the enemy.

In the case of a contact reconnaissance, or of a moving protective force, such as an advanced guard, patrols of this type should be distributed over the whole front. These patrols may be from time to time within view of each other, but the nature of the country or the number of men available will not always admit of this. It is therefore seldom possible, even if it were desirable, to guide a widely extended line of patrols by making them conform to the movements of a directing patrol. Even if adjacent patrols were permanently within sight of each other, any attempt at such a method of direction would result, if well done, in an objectional rigidity, and if badly done, in continual extending and closing, due to the effort to keep the proper intervals. Drill systems of this kind cannot be accurately carried out without dividing the attention which ought to be given wholly to reconnaissance. The direction of a patrol should either be given independently, by map, compass, or landmarks, or the patrols should be informed of the route to be followed by their respective supports, and instructed to keep direction and maintain position, each relatively to the direction and position of its own support.

In spite of the limitations imposed on his initiative, the leader of a contact patrol will find plenty of scope for the exercise of any skill in scouting he may possess. For it must be remembered that one aim of all reconnaissance is to discover the enemy before being seen, and this end alone cannot be achieved with certainty unless every precaution be taken. In contact reconnaissance there is, or there ought to be, but little risk of interception, but owing to the necessity of a regular advance, there is rather greater probability than in independent reconnaissance of unexpectedly meeting the enemy at close quarters. Observation to the flanks is of secondary importance; it is in front that decisive information may be gained, and it is towards a possible enemy in front that the patrol leader should direct his attention.

A contact patrol must advance approximately on a specified line, and it must at least keep pace with the troops which follow it. It is evident, therefore, that from time to time it must advance in the open under the possible view of hostile observers, and that in such case it may at any moment be exposed to fire. There is no evading these risks, nor is there any certain method of guarding against them; the risks must be accepted. But fortunately, they are concurrent; the exposure to fire, which is a danger to the patrol, is dependent on the exposure to observation, which is a danger to the success of the reconnaissance. To lessen the chance of being observed, the best course to pursue is to cross the exposed ground at speed, halting at any convenient point, under cover if cover exists, and if there is no cover, then halting in the open and remaining still. If a regular advance be made, the movement across the open may occupy some time, and during the whole of that time, the moving object will be likely to attract the eye of a hostile observer, while if the movement be made quickly, it may be accomplished while the enemy is looking elsewhere. Even if the patrol is exposed when it halts, it may quite possibly escape notice if it remains motionless. As for the risk to the patrol itself, there is undoubtedly less danger in swift movement than in slow, as an enemy is likely to be disturbed both in nerves and in aim by a rapid advance towards him.

When actually closing on a position possibly held by the enemy, it is advisable to move obliquely towards it, as the target thus presented is more difficult, and the speed need not be checked if retreat should be necessary. When the enemy is met with, the leader of a contact patrol has two equally important duties to perform; he must report to his superior officer, and he must endeavour to keep touch with the enemy, unless the hostile force is inconsiderable and retires immediately. The stronger the enemy, the more necessary is it that he should be continuously observed; therefore, the patrol leader must be resolute. If the enemy appears to be already aware of his presence, or of the presence of any part of the force to which he belongs, it may be incumbent on him to advance still farther, in order to obtain more definite information, and he must continue to conform to the movements of the enemy in front of him, advancing when possible, retiring only when threatened. If the enemy appears to be ignorant of the proximity of the force, the patrol should remain in concealment as long as the enemy remains within view, or until the patrol leader receives further instructions.

The patrols sent out by a stationary protective force usually proceed on definite beats at such intervals of time as the commander may decide. It is desirable that the routes to be followed should be known to the vedettes or sentries of the fixed protective line, in order that the patrols may not be mistaken for the enemy. The system of advancing under cover of darkness and returning by daylight is very suitable for regular patrols of this type. The hours at which patrols are despatched should be varied from day to day, otherwise there is a temptation to the enemy to prepare an ambush.

Protective patrols should not hesitate to engage any party of the enemy which they may meet. If the enemy's force be merely a patrol, a capture may be made, or at least, the enemy's efforts at close reconnaissance may be discouraged. If the hostile force be a large one, then the sooner the outposts are put on the alert by the sound of firing, the better.

CHAPTER VII

THE RECONNAISSANCE OF GROUND

In reconnaissance, the first consideration is usually the enemy; ground, even that which the enemy occupies, is of secondary interest. For although in every theatre of operations there are positions or localities of tactical, and especially of strategical importance, the value of these must always be influenced by the dispositions of the enemy. The mere examination of ground which is not, but may become, the scene of operations, is hardly reconnaissance in a military sense; it is more properly described as military surveying, and those who undertake work of this kind, except in these cases where the inhabitants are so hostile that they may be considered as enemies, should be trained as surveyors, not as scouts. When the ground to be examined is occupied by or is within the possible sphere of action of the enemy, then surveying becomes a branch of reconnaissance;[145] when the enemy and the ground must be examined simultaneously, and their influence on each other considered, the highest skill in reconnaissance is required, for this mutual influence is often of extreme importance.

---

145. As was the case prior to the 6 June 1944, D-Day invasion. On 31 December 1943, a covert soil sampling operation was conducted by the British Royal Navy's No. 1 Combined Operations Pilotage and Beach Reconnaissance Party near Luc-sur-Mer, France (the beach was later code-named Sword). The reconnaissance team's goal was to obtain soil samples of the beach so engineers could determine whether the sand would support the tanks, trucks, and bulldozers that would be part of the assault.

The study of ground in its military aspect maybe conveniently divided into the appreciation of its possibilities and the ability to reproduce its features. Knowledge of the possible influence of ground is one of the first essentials of military training; its importance is impressed on every recruit; its value is demonstrated on every battlefield. A soldier is expected to understand, according to his rank, how to utilise the accidents of ground for the military advantage of himself or of the troops under his command. The scout who is a trained soldier has, therefore, in all probability, knowledge of this kind in some degree. But if his observation of ground is to be of much practical value, his knowledge should not be strictly limited to the requirements of his own rank. He does not gather information for himself, but for his superior, and the more nearly his knowledge attains to the requirements of the superior to whom he reports, the more valuable will his information be.[146] A cavalry trooper, for instance, who has discovered a body of the enemy, should not confine his observation of ground to the selection of a concealed position for himself, and of a suitable line by which to return and report, but should endeavour to note anything which may be of use to his squadron commander; whether the ground is suitable for mounted combat; whether there is room for the squadron or the regiment to manoeuvre; whether dismounted action would offer good results; whether there is cover behind which more of the enemy may be concealed. A soldier who is sufficiently intelligent and well trained to be chosen as a scout ought to be able to give a sound opinion on such points, and this opinion may be of infinite value to his superior. But if the man knows nothing beyond his own

---

146. Henderson is alluding to information that has the potential to influence strategic planning rather than merely operational or tactical decisions.

duty, he will be quite unable to form an opinion, nor, indeed, would he be likely even to think of doing so. In the same way, but in greater degree, if a subaltern[147] on independent reconnaissance should find the enemy in position, the most detailed knowledge of the influence of ground on the manoeuvre of a company will not be sufficient to enable him to appreciate its influence on the manoeuvres of two opposing armies. Yet it is only in this wider sense that his observation of the ground can be of use to his commander, and the more capable he is of considering it from the commander's point of view the better is he equipped for the performance of his duty.

It is unnecessary here to enter on the question of the influence of ground on tactics. It can be studied in the textbooks and worked out on maps; it is brought to notice in every military examination; best of all, it can be practically observed in the field and on manoeuvres. It is a tactical matter, and its relation to reconnaissance is merely that a thorough knowledge of it is one of the military acquirements which are of value to those who may be employed on this duty. There are some who scoff at the idea of subalterns being taught to "think in army-corps," but there may be, as there have been, occasions when momentous decisions have to be made on no better information than is contained in a subaltern's report.[148] If the intelligence of the subaltern has been cramped to such a degree that he has not been allowed to advance beyond "thinking in sections," there will be some uncertainty as to the result of an operation of war initiated on his information. The military knowledge which may be of use

---

147. *Subaltern* is a British military term for a junior officer—a subordinate–below the rank of captain.

148. This point underscores the importance of having officers who are educated; and regarding intelligence work, having analysts who are well qualified academically.

to an officer on reconnaissance has no limit. The wider the knowledge of the informant, the more useful the information.

In cases where the observation of ground is not tactically connected with the disposition of the enemy, where mere topographic detail is required, which the proximity of the enemy makes it difficult or hazardous to obtain, the problem is of a much more limited kind.

Such information is obtained by the application of ordinary methods of reconnaissance, independent or contact, according to the necessities of the case. The measures by which topographic information is acquired are the same as those which are used for obtaining any other kind of information; as a rule, however, it is much less difficult to reconnoitre ground than to reconnoitre the enemy.

Topographic information may be conveyed by maps or sketches or by written or verbal reports. In some cases, no report is necessary, a knowledge of the ground sufficient for the accurate guidance of troops being all that is required. The methods of compiling reports in the form of maps or sketches are fully dealt with in all works on military topography; strict rules, based on practice under peace conditions, have been laid down to assure that no topographic detail shall be omitted, and that the information shall be conveyed in a thoroughly academic manner. Topography has, in fact, become the favourite field of the military pedant; because it can be practised in peace time, it has acquired an exaggerated importance out of all proportion to its value in war; it has even usurped the name of reconnaissance, as if it covered the whole of that art of which it is a minor branch. Ability to depict graphically the features of ground is no doubt a useful accomplishment and is often of assistance in reporting information which has been acquired by reconnaissance, but it is only a part, and not the most important part of the general military knowledge which may be utilised on reconnaissance duty.

The mapping of a region in which the enemy's troops are active must always be a most difficult task; and if an accurate survey be required in a reasonable time, it can hardly be executed except under the protection of an adequate force. The operations of a contact reconnaissance give opportunities for combined survey work, which may result in a satisfactory map of the region traversed; and a record of the route of an independent patrol may sometimes be kept with some accuracy, even if the enemy has to be evaded; but success in such endeavours cannot always be hoped for. Accurate surveying is, in fact, suited only to peace conditions. When the enemy cannot interfere, it is feasible; but it cannot be carried on with any good result if the topographists are exposed to danger or interruption.

Mapping which is frankly inaccurate and pretends only to give, in reasonable proportion, the main features of a district, can nearly always be carried out, and is often of great value. Maps drawn roughly by eye, with perhaps a few compass bearings as guides, or maps sketched in from memory after the ground has been crossed and observed, may, and often do, present exactly the information which is required. The ability to depict ground in this way is the reward of practice in topographic work; and much of the real military value of training in surveying lies in the probability that such training may impart skill in rough sketching. For the trained surveyor is quick to recognise the distinctive features of any piece of country—the watersheds, the direction of the streams, the general trend of the contours; he is also accustomed to drawing to scale and can therefore hope to preserve some proportion in his work. Given the necessary skill, there should be little difficulty in executing such sketches on reconnaissance; but it must be remembered that the value of such work depends entirely on the importance of the particular piece of ground depicted. If a position held by the enemy is being reconnoitred, a rough sketch from eye or memory, showing the approaches to the position, may be of great assistance to the commander, because of the extreme importance of information concerning this particular piece of ground. But it is only in such definite cases that rough work is of value, for it is not sufficiently accurate either for permanent record or for combination with other sketches.

Similar in feasibility and in value to these rough maps are freehand or panoramic sketches. These are particularly useful, when an enemy's position has been reconnoitred, in conveying an idea of the appearance of the visible portion of the ground which is occupied by the enemy. Ability to produce such pictures is dependent to some extent on artistic skill, but some facility can be acquired by practice. The value of freehand sketches, as of rough maps, is entirely dependent on the military value of the ground depicted; unnecessary sketches are frequently added to reports to increase their apparent value, or to hide the lack of useful information. An artistic presentation of an unimportant piece of ground is of no more value than is elegant penmanship in a poor report.

Written or verbal reports do not convey general information about ground in a convenient form but are often sufficient to give enlightenment on definite points. So much is this the case, that if general information be required, and cannot be presented in graphic form, it is usually more satisfactory to obtain it from the observer by personal questioning than to receive it in the form of a report. The method is that usually pursued when extracting topographic information from the inhabitants of a country. By a succession of definite questions on definite points, the information required may frequently be obtained from persons who would be quite unable to give a general description of the country. As a rule, however, when topographic information is required from officers and men employed on reconnaissance, it is not of a general nature; their attention is usually directed to some particular and definite objective, and in many cases a written or verbal report is all that can be expected or desired.

Officers on reconnaissance are frequently able to acquire a good deal of topographic information from friendly or neutral inhabitants. The amount and the accuracy of this information depends very much on the knowledge of the questioner, as to the manners and customs of the people whom he examines. For features of ground do not present the same aspect to different men: a carrier takes no interest in any hills except road gradients; the driver of a locomotive will probably be able to place correctly the stations on a hundred miles of railway, and will know nothing of the country a hundred yards from his line; a shepherd may have a thorough knowledge of a large tract of country, but is likely to be vague as to distances. A useful method of procedure is to find out what the informant's business is, to question him on matters that are likely to be within his knowledge, and to consider the answers, as far as possible, from his point of view. For example, if a carrier, or transport rider, were to say that a river could be crossed anywhere, he might quite possibly mean that there were sufficient bridges and fords for his purposes, and not that the river was passable at any point. The accuracy of the description "a good road," depends entirely on the standard of the roads to which the informant is accustomed.[149]

In many countries the length of a conventional "day's journey" and the distance represented by "an hour" must be

---

149. This example highlights the subjectivity of descriptors—like "good"—rather than reporting the physical condition (i.e., the facts). A corollary to this is Henderson's recurring theme for the need of commanders to ask the right questions.

known if distances are to be accurately interpreted. In civilised countries distances can usually be ascertained from maps, but among uncivilised races, or in unsurveyed countries, methods of reckoning distance vary considerably, and estimates require a good deal of corroboration before they can be accepted. Even descriptions of particular features of country, or of communications, sometimes require to be skilfully interpreted to be of any value. A scout in South Africa was questioned as to the practicability of a mountain path. "Well," he said, "it isn't very good, but I wouldn't mind riding down it on a borrowed horse." What he might have meant was that it was so good that any horse could get down safely; what he did mean was that it was so bad that he would not risk his own horse on it.

In dealing with persons who are willing to give information, the utmost patience and good humour are required, for very often it is only in a roundabout way that anything comprehensible can be elicited. When useful information has been obtained and even corroborated, it must still be remembered that second-hand information is not the best. It is useful as a guide to further reconnaissance; it is admissible, when reported as hearsay, in relation to matters which cannot be learned by personal reconnaissance; but in no case should it be offered or accepted as a substitute for information which could reasonably be obtained by observation.

CHAPTER VIII
THE TRANSMISSION OF INFORMATION

The aim of all reconnaissance is the acquisition of information, and it is in this pursuit of information that the real difficulties of reconnaissance are encountered. But in order that the best use may be made of the information which is obtained, it is necessary to ensure that every item of intelligence reaches the person who is able to utilise it to the greatest advantage.[150] This can only be done by a proper chain and system of transmission; it is but seldom that the man who actually gains the information can select the particular officer to whom it will be most valuable. As a rule, the scout cannot be expected to do more than pass his information to his own officer, or to some other superior who may be in a position to transmit it to his own officer.[151] In certain exceptional cases,[152] an intelligent scout may judge that his information is of urgent importance to some particular officer or body of troops, and may, on his own initiative, take measures to communicate his news directly; but responsibility for such independent action ought not to be imposed on scouts, save with reference to the most

---

150. In the terminology of intelligence, these people are known as *customers*.

151. i.e., through the unit's chain-of-command.

152. Note the reference to the adjective, *exceptional*. In military and other hierarchal organisations that have a chain-of-command structure, such cases do need to be exceptional. This means there must be little doubt that any other decision would end in disaster. As Henderson points out later in this sentence, a surprise attack.

unmistakable cases, such as giving warning of an unexpected advance of the enemy.

The first and indispensable method of transmitting information is by the ordinary channel, from inferior to superior. Thus, a cavalry scout engaged on contact reconnaissance reports to his patrol leader, the patrol to the squadron, the squadron to the regiment, the regiment to the brigade.[153] The officers through whom this report passes must, each in turn, judge whether the information affects any troops other than those under their immediate command.[154] If it does not, the report is retained; if it does, it is recorded and passed on to the higher authority. It is necessary to record it, not only to safeguard the officer, should the information be lost in transit, but also to enable him to keep abreast of the situation, so that he may be able to gauge the value and accuracy of later reports.

The responsibility for this first transmission from lower to higher authority ought to rest on the unit by which the information is first obtained. If two units acquire the same information independently, it would, of course, reach the higher authority through two channels, and the separate reports would be corroborative.[155] But if, at any stage of its progress, a report is passed from one unit, or one channel, to another, the recipient must be told whether he also is expected to transmit it, in order that the same information may not reach higher authority through

---

153. i.e., via the chain-of-command.

154. What he is discussing is the need-to-know and the need-to-share doctrine that is at an axiom of intelligence work.

155. This is a serious issue, because self-validating reports can lead a commander to make decisions that have grave consequences. In this context, self-validation reports are created by pieces of information that give undue weight to the data, or from the same source but appear to have come from independent sources.

two channels, as if the reports were independent, instead of being merely the same report duplicated. To guard against such unnecessary and perhaps misleading duplication, some further arrangements are necessary.

If the case be considered of two regiments, 25th Hussars and 26th Lancers, engaged on contact reconnaissance, and if it be supposed that a scout from the inner squadron of the 25th Hussars discovers the enemy in front, the normal progress of the report will make the matter clear. The scout reports to the patrol leader; the patrol leader reports to the squadron commander. So far, all is simple; but the squadron commander has a double responsibility. It is necessary that the information should be passed to brigade headquarters, but it is necessary also that the other regiment should be warned. The squadron commander therefore duplicates the report; he sends one copy to the Officer Commanding 25th Hussars, his own regiment, and the other to the adjoining squadron commander of the 26th Lancers, and at the end of both reports he writes "addressed O.C.[156] 25th Hussars, repeated O.C. left squadron 26th Lancers." The first address gives the main channel of communication, in which every recipient of the report is responsible for the transmission of all relevant information to higher authority; the second address gives the branch line, in which the recipient is responsible only for the enlightenment of those whom he considers directly affected by the information. Each time the report is transmitted on this branch line, it should bear a notification to the effect that it has already been forwarded to headquarters through O.C. 25th Hussars.

When this method is adhered to, the responsibility for the transmission of information is definitely fixed, and any failure in

---

156. O.C. is the British military abbreviation for Officer Commanding.

communication can be brought home to the responsible person. If it be not adhered to, there is a prospect of the means of communication of a whole force being temporarily congested by the transmission to the commander of copies of the same, perhaps erroneous, report, through half-a-dozen different channels. Not only is this a waste of work and of valuable time, but it is almost inevitable that the commander will be deceived by the succession of similar reports, and in the belief that some of them are based on independent observation, will give undue weight to the information.

The short-cut for passing information from one cavalry unit to another has here been taken as squadron to squadron, and, as a rule, especially in contact reconnaissance, this route is convenient. But it is within the discretion of any officer to transmit intelligence to neighbouring units, for reasons either of urgency or of convenience. The urgency he must decide for himself, remembering always that should he not do so, each of his superiors in succession, will have an opportunity of passing the information to other units, and that finally the information will reach the common commander, who can distribute to his subordinates such items as he thinks necessary. With regard to convenience, the subordinate may know that he is in closer touch with neighbouring units than his superior officer is, and in such case, he may well utilise this proximity to transmit his information at once.

Although the direct channel of communication from inferior to superior of the same unit must always in the end be adhered to, it may frequently happen that communication by this channel may be slow and difficult, and that there may be a certainty of rapid communication by another route. This is often the case when information is obtained by independent reconnaissance and is not uncommon even when contact methods are employed. An independent patrol, for instance, returning to headquarters on tired horses, may meet with a body of their own troops. The patrol leader should at once hand over a report "for transmission"; the officer who receives it immediately becomes responsible for the transmission of the information not only to the person to whom it is addressed, but also, if it be of value, to his own superior. Thus, if the operations of the 25th Hussars and 26th Lancers be again considered, and if it be supposed that an independent patrol from the 25th Hussars, returning, meets a contact patrol of the 26th Lancers, the transmission of the information might properly proceed in this way. The contact patrol commander would either receive the report himself and send it to his squadron commander or would detail a man to guide the independent patrol to the squadron commander. In most cases, the latter would be the better course, as a contract patrol can seldom afford to delay. The squadron commander receiving the report would then decide how to transmit it. If he has good communication with his own regimental headquarters, he would probably send it on as it stands to O.C. 26th Lancers, endorsed: "Received from Lieut. X., 25th Hussars, for transmission to O.C. 25th Hussars." The O.C. 26th Lancers, when he receives it, must decide whether, considering the circumstances of the case, it will be sufficient for him to send the report to O.C. 25th Hussars only, or whether he should duplicate the message and send a copy direct to the brigadier. If he adopts the latter course, the report sent to O.C. 25th Hussars would be endorsed: "Copy sent to brigadier," and the copy to the brigadier would be marked: "Original sent to O.C. 25th Hussars." The O.C. 25th Hussars is thus cut out of the channel, and, when he receives the report, has no responsibility except to pass the information to the troops under his own command. The necessary check and safeguard which provides against any breakdown on this loop line is supplied by the leader of the original independent patrol, whose task is not ended until he has reported personally to the officer who sent him out, or has received fresh instructions from that officer, with an intimation that his report has been received.

It may sometimes occur that the information to be transmitted is of such importance or urgency that an officer is justified in sending it direct to the commander-in-chief, as well as to his own superior. But if communication arrangements are properly organised, such direct transmission is probably no more speedy than is transmission through the ordinary channel. In any case, the repetition of a message direct to headquarters is an exceptional measure, and an officer should not adopt it unless he is firmly convinced of its urgent necessity.

In order that intelligence reports may be clearly understood, and their significance appreciated by those who receive them, it is desirable that every report should have on it a number, the date, and the name of the place where it is written, or from which it is despatched. The importance of the date and of the place is evident, and our official textbooks contain strict regulations on this subject. The value of consecutive numbering is not, however, so generally recognised, except as a convenience in the matter of acknowledging receipt or asking for further particulars. Yet numbering, if carried out properly, is more than a convenience: it is a safeguard in establishing continuity, either in reports or in instructions; but to be of any value, it must proceed on a recognised system. One such system is that of the double series, which may be briefly explained. Each officer who may have to send messages of any kind uses two series of consecutive numbers, one series bearing a distinguishing letter. All messages sent to his immediate superior are numbered from this lettered series; all other messages are numbered from the other series. The reports which reach the superior ought therefore to bear consecutive numbers, as, G.1, G.2, G.3, etc. If there should be a gap in the numbers, it is at once known that a report is missing, and inquiry can be made. If all messages be numbered on a single series, the superior will never be in a position to know whether a message is missing, unless it should happen that the officer reporting has sent no communication to any person except this superior. There is no extra trouble entailed by using this double series, and its advantages are considerable.

These few principles, if intelligently considered and practised in peace, ought to be sufficient to ensure the proper transmission of information in, and from units, in war. Their principal object is to impose a definite responsibility on all regimental officers in this respect, so that information may not be lost, or delayed in transit. Every officer of every branch of the service should have this responsibility impressed on him and should be instructed and practised in this duty. The necessity for such instruction has seldom been recognised, and, as a consequence, the loss of information in most wars has been excessive, and not infrequently disastrous. Common-sense is the basis of these principles, but training is required in order to ensure the swift and unfailing application of common-sense in harassing and dangerous circumstances. A simple and definite system of procedure is a valuable aid when mind and body are tired, strained, or overworked; and these conditions are frequent in war. And if a system is an aid, not less is personal responsibility a spur unequalled in efficacy.

The transmission of information between the higher units, brigades, divisions, armies, is carried out by officers of the general staff. The information acquired by reconnaissance is, of course, only a fraction of the information received from all sources; and the study of the methods by which the latter is considered, collated, forwarded, and distributed is a matter of intelligence rather than of reconnaissance work. The principles on which transmission is arranged may, however, be briefly stated; for if there should be any failure in this transmission, the best reconnaissance will be of little value.

There are four principal directions in which a general staff officer must distribute information—his own general, the immediate superior of his own general, the troops belonging to his general's command, and other troops. The first duty is obvious; it implies the collation and recording, for the use of the commander, of all information which reaches the command. It is a continuous duty, for information may arrive at any time; if it does not arrive, the staff officer must go in search of it. Nor is simple collation and recording sufficient; the information must be studied and considered; efforts must be made to verify doubtful reports, and the directions in which further information may probably be obtained must be indicated to officers and men engaged on reconnaissance or other intelligence duties. The general staff officer is, in fact, responsible to his commander for information.

The forwarding or distribution of information by the other three channels which have been mentioned, is carried out by a general staff officer in the name and with the authority of his general. With regard to the transmission of information to the general's immediate superior, whether he be the commander-in-chief or an intermediate commander, the rule is simple: all information which may be of use to him must be sent on, unless it is known to be already in his possession. No rule less stringent will fulfil the requirements of efficient intelligence service.[157]

The distribution of information to the troops of the command itself is to ensure that all subordinates are kept abreast of the situation, in order that they may be able to understand the bearing of orders or instructions which may be sent them, and that in the absence of orders they may, if emergency arises be

---

157. This advice relates to many so-called intelligence failures, whether through the direct, indirect, or consequential lack of sharing information.

able to take action on their own initiative, and cooperate to the best advantage.

The passing of information to other troops is not invariably the rule; it is usually confined to cases of urgency, or to situations where communication between two bodies of troops is easier than is communication between either of them and their common commander. In the case, for instance, of two brigades engaged in operations, it is, as a rule, better for a brigadier to forward his information to divisional headquarters only, thus leaving it to the discretion of the divisional commander to pass to the other brigadier such portions of this intelligence as he may think necessary. But should communication with divisional headquarters be slow or uncertain, or should a sudden crisis occur likely to affect both brigades, the brigadiers would of course endeavour to transmit information direct to each other.

Information may be reported either verbally or in writing, and there are certain advantages in each method. The first report, made by the person who actually discovered the information, is likely to be better understood and appreciated if delivered personally, by word of mouth. For thus the essential part can be separated from unimportant or irrelevant detail, with no loss of time; accuracy can be assisted by questioning, and there should be no possibility of misunderstanding. If, however, the information is to be transmitted, it should be written; even if it be decided to send on the original informant, in case the superior officer should desire to interview him personally, the written notes of his report should accompany him, or be otherwise forwarded. There is then no danger of serious discrepancies or misconceptions, such as are common when verbal reports are repeated. The principal objection, however, to the verbal transmission of reports is the absence of any record. Whatever is worth transmitting, is worth writing down, and the rule requiring that messages for signals, telegraphs, or telephones shall be written should always be adhered to.

The information contained in a report should be relevant and accurate, and it should be conveyed clearly and briefly. That is, the information should be confined to matters which may conceivably be of military value, and these should be given with the most scrupulous exactness. Definite statements are desirable, but care must be taken to make no definite statements unless there is definite knowledge behind them. Definite statements on insufficient authority are misleading and dangerous.[158] If the strength of the enemy, for instance, has been roughly estimated, it should be so stated; if it is reported on hearsay, the authority for the statement and an opinion of the credibility of the informant should be given.[159] And it is not enough that a statement should be accurate: it should convey an accurate impression. If a scout is fired on from a certain hill, that is not in itself sufficient to justify him in reporting that the hill was held by the enemy. The enemy may have been there quite fortuitously, with no intention of holding the hill, and may have fired merely for his own security. What the scout is justified in reporting is the exact truth, that he was fired on.[160] If he has

---

158. In the language of intelligence, *probability*.

159. According to Prunckun, "This system is known by various names by the American, British, Canadian, Australian and New Zealand militaries, including 'source and reliability matrix,' NATO System, Admiralty Code, but are based on the British World War II so-called *Admiralty Grading System*." See *Methods of Inquiry for Intelligence Analysis, Third Edition* (Lanham, MD: Rowman & Littlefield, 2019), 44–45.

160. This is a poignant example of where an untrained observer tries to interpret the information (using inferential reasoning), rather than just transmitting the facts so an analyst can interpret them in the context of other data using structured analytic techniques.

carried his reconnaissance farther and has actually ascertained that the hill is occupied for defence, he may then accurately describe it as held, because the impression which will be conveyed by his statement is accurate. His whole endeavour should be to enlighten his superior, and technical accuracy is no excuse for making a statement which is open to misconstruction.

The relevance, which in reconnaissance is almost the same as the military value of a report, is a matter which cannot always be within the scout's knowledge. Military value is dependent on the situation, and perhaps on the magnitude of the operations; information which is valuable to the patrol leader may have no interest for the brigadier. If the case be imagined of a contact reconnaissance on a large scale, already in contact with the enemy, and of a single scout being chased by a troop of the enemy's cavalry, the scout's report of his adventure would be of value to his patrol leader, and perhaps to the squadron commander, but would be of no value to the commander-in-chief of the army. But if it should happen that the scout should recognise this troop of hostile cavalry as forming part of a particular regiment, this information might be of the greatest value to the commander-in-chief himself, for it might enlighten him as to the composition, and even as to the strength, of the troops opposed to him. Therefore, every officer should consider how much of a report is of value to himself alone, and how much is worth transmitting to his superior.[161]

---

161. Although sound advice, Henderson has overlooked what is termed *filtering*. That is, if each officer in the chain-of-command can decide what information is likely to be relevant to his or her superior, the consequences of this repeated filtering does not have to be spelled out to understand. Incomplete information can have disastrous consequences for tactical decision making. This issue emphasises the doctrine of "need-to-share."

It is not always judicious to be very strict with scouts as to the relevance of their reports. With some men, otherwise capable enough, the personal interest of their duty rather obscures the larger issues, and they are unable to impart their information unless they are permitted to relate their adventures and illustrate their exceptional ability. But irrelevant matter of this kind should not survive the first transmission of the report; the recipient of any information should be able to retain an attitude sufficiently detached to ensure the suppression of unnecessary details.[162] Irrelevance may be excused in the finder, never in the transmitter.

Clearness and brevity in a report are virtues which are difficult to combine.[163] For sometimes brevity is sacrificed in the endeavour to avoid the possibility of misunderstanding, and reports become as long-winded as legal documents. Still more frequently reports are made so brief as to be incomprehensible, and become not only valueless, but exasperating.[164] Clearness must always be the essential; if brevity can be combined with it, so much the better; but it is better to be tedious than to be obscure.

The consideration of this question of reporting information leads inevitably to the recognition of the importance of general military knowledge, not only in those officers to whom the search for information is entrusted, but also in those through

---

162. A good point, but what constitutes *unnecessary* in intelligence work can be a fine line, because to some degree, all information can be relevant.

163. See James S. Major, *Writing Classified and Unclassified Papers for National Security* (Lanham, MD: Scarecrow Press, 2009).

164. Here he is acknowledging that lack of detail can lead to confusion, hence training in report writing is important (which includes oral briefing and graphical presentations).

whom information is normally forwarded. For the final value of reconnaissance depends very much on the ability of junior officers to discern the facts or deductions which will be of use to their superiors. Such discernment is partially, no doubt, dependent on natural judgment and sense of proportion, but it is chiefly acquired by sound military education. The officer who has attained only to the standard known as the "good regimental officer," is insufficiently equipped for reconnaissance work. Tactical or administrative knowledge of a single unit, or of a portion of a unit, is of very limited value when information of the enemy is required. What is of value is a knowledge, in some degree, of the art of war.

CHAPTER IX

AERIAL RECONNAISSANCE

The advent of the dirigible and the aeroplane has undoubtedly introduced a new factor into reconnaissance which must alter considerably the methods by which information of the enemy is acquired in war. What the final effect may be is not yet known, for the art of flying is only beginning to develop; aerial reconnaissance has not yet been seriously tried in war, and theories based on peace experience are not quite trustworthy. The art of flying, however, wiping out as it does both the obstacles of nature and the obstruction of human opponents, has so enlarged the scope of reconnaissance that it is not difficult, even now, to recognise certain definite extensions of the possibilities of gaining information, which must modify to some extent the methods hitherto adopted in warfare. There is nothing in the new art, surprising as its development has been, to lead to the belief that it will cause any modification of the principles of war; it is in the application of the principles, in the methods of warfare, that new influences will have their effect, and within these limitations the effect is likely to be profound.[165]

It may be advisable to consider first what the possibilities of flight are, at the present moment, with reference to its employment as an aid to the acquisition of information in war. The large dirigible has a radius of action of some 500 miles; that

---

165. And profound they were! One just needs to think back to how aerial reconnaissance developed from cumbersome handheld cameras used by air crew in the First World War to appreciate the sophisticated equipment that features in today's reconnaissance satellites.

is, it can fly continuously for 1,000 miles at a speed of over 50 miles an hour.[166] This speed is sufficient to enable it to progress against winds of normal velocity, and its range is more than sufficient for all purposes of land reconnaissance. But the dirigible of this type is a very delicate and a very vulnerable machine; it requires a large shed to house it, and very careful handling on starting and landing. In flight it is a good target and can be destroyed or damaged by gun or rifle fire from the ground, or from other aircraft. Its great advantage is that it can navigate the air by night, and make such observations as are possible in darkness, in leisurely fashion, without being exposed to any serious danger from hostile action, except from another airship. As against an army in the field, the airship has, as yet, no very great value as an offensive weapon; its powers of bomb-dropping or of fire action against troops or war material in the field have not yet developed beyond the ability to cause annoyance and comparatively unimportant damage.

For such warfare as the British Army is accustomed to wage, the use of the large dirigible presents many difficulties. Continental nations, with coterminous frontiers, can build their airship sheds in positions strategically suitable for reconnaissance of the theatre of war in which their armies will most probably operate. Our scattered Empire[167] and our sea frontiers do not lend themselves to such arrangements, except for

---

166. Five-hundred miles converts to 805 kilometers; 1,000 miles about 1,609 kilometers; and 50 miles per hour is about 80 kilometers per hour.

167. It was a period that historians refer to as Britain's "imperial century," because there were an estimated 10,000,000 square miles (26,000,000 square kilometers) of territory and approximately 400 million people that comprised the British Empire. See Ronald Hyam, *Britain's Imperial Century, 1815–1914: A Study of Empire and Expansion, Third Edition* (New York: Palgrave Macmillan, 2002).

naval reconnaissance. An airship base in this country would be too far from the scene of any European conflict in which our troops would be likely to take part, and the erection of a large airship shed at an oversea base is an undertaking which would require a considerable time for its completion; so much time, indeed, that the campaign might well be over before the shed was completed. A type of dirigible may be evolved, not improbably, which may be able to live through bad weather without the protection of a shed; to ride out gales, anchored to the ground, or moored to a post, while still competing with the large rigid airship in speed and range of action. Such a craft would be of value to us for warfare in Europe, and of even greater value in our more frequent wars against the peoples who dwell near, or in, our distant possessions. For the present, however, it is unlikely that we shall be able to utilise, in either case, any dirigibles except those of the small, "portable" type, limited in speed, in range, and in weight-carrying capacity, inferior to the large airship in all military qualifications except manoeuvring power and durability.

The aeroplane has a shorter, but quite effective radius of action of some 200 miles, and a machine which is safe to land in ordinary country may have a speed of over 90 miles an hour.[168] No doubt both range and speed will soon be increased,[169] but for purposes of reconnaissance in land warfare such increases will not much affect the conditions under which flying must be undertaken. The aeroplane is a very difficult target either from

---

168. Two-hundred miles is approximately 322 kilometers and 90 miles per hour is about 145 kilometers per hour.

169. And it did; take for instance Lockheed Aircraft Corporation's SR-71 that had a service ceiling of about 85,000 feet and a reported cruising speed of Mach 3.2. Once fully fueled, the SR-71 claimed it could fly some 2,500 miles before it needed to be refueled.

the ground or from another aircraft, and there is no difficulty in providing armoured protection for the pilot and observer, and for the engine. Nor is the machine itself very vulnerable to small-calibre bullets or to shrapnel; it may be frequently hit without suffering vital damage.

It is unnecessary to enter further into the details of the flying machines which may be used in war, except that one point must be made clear. No aircraft has yet been produced which would be able absolutely to prevent a hostile aeroplane from carrying out a reconnaissance.[170] There are, of course, crude methods of interfering with another aeroplane, such as "giving him the wash"; that is, crossing the adversary in such a manner that he may be hampered or upset by the slip-stream of the screw; light guns or machine guns may be mounted in fighting aeroplanes; rifles or pistols may be used by passengers or pilots; endeavours may be made to get above an enemy and throw bombs at him, or in the last resort to ensure his destruction at the price of self-immolation by ramming in mid-air. But to employ any of these methods, it is necessary to get within short range of the adversary, and herein lies the difficulty of bringing him to action. An observer in the air finds it difficult to see another aeroplane, unless it is either well above him or well below him; to a gunner in the air, another aeroplane presents a very difficult target, except at the closest range; and to strike, or "give the wash" to, an efficient aeroplane which is trying to evade him,

---

170. Despite the advancements in aircraft technology, weapons systems to engage reconnaissance aircraft have kept pace. As an example, Lockheed's U-2 was able to overfly the former Soviet Union for four years (July 1956 to May 1960) before the Soviets developed a surface to air missile that could bring down this ultra-high-flyer. A Soviet surface to air missile was also responsible for bringing down a U-2 over Cuba during the Cuban Missile Crisis in October 1962.

means not only a close approach, for which superior speed and manoeuvring power are required, but a definite superiority of skill and resolution on, the part of the pilot. The speed of aeroplanes is so great that, except in the case of long strategical reconnaissances (for which, moreover, only the fastest machines would be used), there is no time for a long chase; an enemy can very quickly return to the safety of his own lines. Until a fighting machine has been produced which will equal the unarmed aeroplane in speed, in climbing power, and in handiness, there can be little doubt that a resolute enemy with a reasonable air force at his disposal will be able to get such information as aerial reconnaissance can acquire, despite our endeavours to stop him.[171] Possibly, by continual interference and by the display of superior resolution, the enemy's courage may be worn down, and something like the mastery of the air secured, but this effect will take time to produce. At the outset of a civilised war, the information obtainable by aircraft will undoubtedly be open to both sides. This is, perhaps, the most important of the new considerations which have to be faced. Under ordinary circumstances the strategic deployment of both armies cannot be concealed. From the time that troops leave ship or railway, they will be under observation. If they move by day, the most stringent precautions will not suffice to conceal them entirely from aerial observation, and such precautions are certain to cause delay. Even if they move by night, large forces are not easy to conceal by day; in billet or bivouac troops are visible, and transport is obtrusive. Aerial scouts will no doubt make mistakes; troops may be reported in places where there are none;

---

171. For instance, once the Soviets developed a surface to air missile that could intercept the U-2, the United States developed the SR-71. With a top speed that exceeded Mach 3, the SR-71 was able to increase speed and simply outrun a surface to air missile.

troops will in other places remain undiscovered. But it must be expected that the principal dispositions and movements will be disclosed, and that the fog which formerly obscured the initial strategic design will be cleared away.

That this enlightenment of both adversaries will have an effect on strategical methods is certain, but what the effect will be is not so clear. Most probably it will cause commanders to be more cautious in their initial dispositions, for unless one commander is certain that he can seize and keep the initiative, forcing his enemy to conform to his movements, both will have the unpleasant consciousness that their weak points as well as their strong ones are known, and that a concentration for offensive purposes at one place may leave another point open to concentrated attack by the enemy. It would seem that strategic surprise would be almost eliminated from the possibilities of civilised warfare in so far as such surprise is dependent on concealment.[172] Only by swift movement, calculated accurately on considerations of space and time, and worked out for the enemy as well as for his own force, can a commander hope to surprise or even to anticipate an efficient adversary in the strategy of the future.

To prove that this effect of aerial reconnaissance is a serious one, it is necessary only to consider the strategy of any great campaign of the last hundred and fifty years. If the strategy of the campaigns of 1800, 1805, 1815, 1849, 1862,[173] 1866, for instance, be reviewed in the light of the development of aerial

---

172. And, "concealment" can take more forms that just hiding—it can include many deceptive tactics. These tactics are covered in detail in Hank Prunckun, *Counterintelligence Theory and Practice, Second Edition* (Lanham, MD: Rowman & Littlefield, 2019).

173. These maps appear on pages 178–184 of this book.

reconnaissance, it becomes evident that the methods even of evolving plans of campaign must be profoundly affected.

The influence of aerial reconnaissance on the accepted system of employing independent cavalry is another important point to be considered. In past wars, strategic exploration has been one of the chief duties of cavalry, and of late years theories and speculations as to the proper method of carrying out this duty have crystallised into a system which is set forth in the textbooks of most nations; set forth, indeed, rather more definitely and dogmatically than the experience of war would seem to justify. Briefly, it may be said that to the independent or strategic cavalry four duties have been allotted:

1. To find the enemy's cavalry.

2. To beat it.

3. To find the enemy's main bodies.

4. To keep touch with them.

It seems probable that, in future, the first and third of these duties will be carried out by aircraft; and there is reason to believe that it may be advisable to await the result of aerial reconnaissance before despatching the cavalry on any errand whatever. It seems unreasonable to send off the main body of the cavalry, a force of extreme value for both tactical and protective purposes, without at least giving them the advantage of the information which may be expected from the first expeditions of the aircraft. This first information may even be so complete that a commander might very well determine to dispense altogether with cavalry strategic exploration, and to devote his mounted force to tactical reconnaissance and tactical action. Moreover, the difficulty, due to aircraft, of concealing the disposition of troops must add greatly to the value of swift moving cavalry for tactical purposes, such as unexpected flank attacks, or the sudden reinforcement of a particular point in the line of battle. The observation of aircraft will render tactical as well as strategical surprise more difficult, and the prospect of being able to utilise a strong and intact mounted force on the battlefield will not encourage a commander to risk his cavalry on the quest for information which is almost bound to reach him from other sources.

The protective duties of cavalry, and indeed of other troops, will also be lightened considerably by the assistance of aircraft, especially by day, except when in close contact with the enemy. For it ought to be impossible for any serious concentration of the enemy's troops within striking distance to escape observation.

On the battlefield, it is probable that aerial observation will inform both commanders with some accuracy of the disposition of their adversary's troops, and where large numbers are engaged there ought also to be warning of decisive movements. It must be recognised, indeed, that throughout a campaign where both sides are sufficiently equipped with aircraft, the game must be played with the cards on the table. It would seem that, in order to gain advantages, night operations will be largely used; accurate calculations of time and space will be of great importance; resolute decisions and swift movements will be necessary. Dispersion of forces will be dangerous, and strategical or tactical errors will be difficult to redeem, as they will quickly come to the knowledge of the enemy.

It is hardly necessary to discuss at length the effect that will be produced if only one side is equipped with aircraft, or if one side can establish an effective command of the air. The advantages are evident, and they may well be decisive.[174] The

---

174. An insightful observation that has proven correct in many military engagements.

principal value of aerial reconnaissance lies in the distance which can be covered; in the speed with which the objective can be reached, and the information brought back; in the fact that there are no obstacles to be met with except hostile aircraft; and in the consideration that the enemy's dispositions in depth as well as in front can be observed.

Illustrations of range and speed can easily be given. Under normal circumstances in war, it ought to be possible to make a fairly detailed observation of an enemy's force, at a distance of three days' march, in about three hours, from the time at which the order is issued until the information is delivered at headquarters, and this would be quite an ordinary reconnaissance. If it were necessary merely to verify the presence of a large force of the enemy at a particular spot, a much greater distance could be covered at a greater speed. The more detailed the observations, of course, the longer time must be spent over them.

Conditions of weather make a good deal of difference both in the time taken for journeys by aeroplane and in the distance, which can be traversed. A great advance has been and is being made in the art of flying in strong winds, and as machines improve it will be only in case of a severe gale that flying will be impossible. But in cases of ordinary reconnaissance, in which an aeroplane must make an "out and home" journey, wind always hinders more than it helps; and when the velocity of the wind approaches the speed of the aeroplane, the delaying effect becomes serious. This reduction of speed due to wind affects also the radius of action, which is dependent on fuel consumption, and therefore on time; a strong wind from any direction consequently affects both range and speed adversely.

The absence of obstacles, natural or of human construction, differentiates aerial reconnaissance from that on the ground.

High mountains may cause difficulty, but no mountain range is impassable to an aeroplane. Forts, barricades, broken bridges present no difficulties, and armed opposition in the air is not yet a serious factor.[175] The difficulties of aerial reconnaissance are those due to the imperfections of the machine, not to any outside influences.

The fact that the whole of the enemy's dispositions are exposed to aerial observation, and not merely the outer fringe of his protective troops, is perhaps the most important difference between the new reconnaissance and the old. The aerial scout can see what Wellington always wanted to see, "what is on the other side of the hill." The occupation of false positions, the secret massing of reserves, demonstrations against pretended objectives, will not in future have the same prospect of deceiving an adversary as in the past. The strength of every column may be ascertained, and its direction of march may be observed. The possibilities of surprise and stratagem will in fact be very much limited.

The height at which aircraft should fly when observing the enemy is a point that has been much discussed, and in making his choice of possible altitudes, the observer has sometimes to consider conflicting requirements. In clear weather and over ordinary country, observation is possible up to seven or eight thousand feet, and it is considered at present that there is but little risk of damage due to fire from the ground at heights over three thousand feet.[176] But when the weather is cloudy or perhaps

---

175. As the twentieth century progressed, and surface to air missiles became a problem, reconnaissance satellites were developed to get around this issue.

176. As anti-aircraft weapons systems matured, this celling had to increase to 70,000 feet—the reported height at which CIA pilot Francis Gary Powers' (1929–1977) U-2 was shot down on May 1, 1960.

when the country is heavily wooded, it may be impossible to observe at three thousand, or even two thousand feet, and a descent to lower altitudes becomes necessary. In such case the observer has to solve the problem which sooner or later confronts every scout: that is, he has to obtain information, but any information he obtains is useless unless he brings it back.[177] His decision must be governed partly by his instructions, and partly by the circumstances of the moment. If a commander should notify that certain required information is essential and urgent,[178] then heavy risks must be faced, for the scout will know that others will be sent on the same errand, and that there is a good prospect of one at least being able to return to his lines. Cases like this, however, are not likely to occur often, for pilots and aircraft are expensive, and not easy to replace, and commanders will hesitate to expose them to more than ordinary hazards, except for a vital object. If it should be necessary to incur such risks, pilots must do their best to minimise them by such means as taking cover temporarily behind clouds, and appearing unexpectedly, or pursuing a devious course, both horizontally and vertically. The same stratagems should be employed to lessen the risks which are inseparable from even ordinary reconnaissance at an altitude lower than the normal; when only general information is required, descents of short duration, during which the aircraft will be visible for quite limited periods, may suffice for the purposes of the reconnaissance. The aerial scout, in fact, like any other scout, must learn to combine caution

177. Today, images of action as it unfolds are available for a commander's immediate viewing regardless of where he or she is in the world.

178. This is termed an *intelligence requirement* (IR), or an *intelligence collection requirement* (ICR). But, in some intelligence agencies, such as those associated with the military, it is referred to as *essential elements of intelligence* (EEI).

with boldness; the more valuable the information that is sought for, the more necessary it is that it should be acquired, and safely communicated. There is, as yet, no very direct evidence on which the vulnerability of aeroplanes to fire from the ground can be accurately calculated. On the few occasions on which aeroplanes have been used in war, a percentage of them have been hit, both by rifle and shrapnel bullets. The extent to which they were exposed to fire is not, however, known; it does not appear that any great volume of fire was ever directed against them. Experiments are now being conducted in several countries in order to ascertain the best method of directing fire action against aircraft.[179]

If[180] aeroplanes become effective fighting machines, capable of bringing each other to action, and seriously interrupting hostile reconnaissance, then of course a new risk will appear which must be faced under all circumstances. But on this matter, there is not much basis as yet to form a theory, or even a reasonable speculation.[181]

Under our normal system of reconnaissance, an aeroplane carries both a pilot and an observer. For long distance, strategical scouting, where it is necessary only to locate large bodies of troops, the single seater will probably be used a good deal, the machine being designed for speed and fuel-carrying

---

179. The history of battlefield anti-aircraft artillery continued from the basic searchlight and small field pieces that were fitted to makeshift mounts and aimed at the sky (which Henderson witnessed), to the shoulder launched guided anti-aircraft missiles of today (i.e., man-portable air-defense systems, abbreviated as either MANPADS or MPADS).

180. Perhaps he should have been less cautious and stated, "when."

181. Although Henderson's predictions were, in the main, insightful, he should have been more confident in his prediction about what was to evolve, especially over the next few decades.

capacity in order to give it long range.[182]   When an observer is carried, usually the pilot finds, the way, and the observer looks out for signs of the enemy.   In bad weather it may be advisable for the observer to set the course, in addition to his other duties, in order to leave the pilot free to manage the machine.

Aircraft will probably be of much assistance to artillery in indicating concealed targets, and in observation of ranging and fire effect.[183]   In this part of aerial work considerable progress has been made, and the results are promising.   The location of hostile troops and batteries is of especial importance.

In this new art of aerial reconnaissance, it is noticeable that the training required for an observer is a much more simple matter than that required for a scout on the ground.   If the observer be carried as a passenger, then all he requires in addition to his general military knowledge, is sufficient experience of the air to enable him to follow a map accurately, and to recognise troops, guns, and wagons on the ground and estimate their strength.[184]   In time, of course, the acquisition of this experience will form part of the ordinary training of a staff officer and it is not improbable that, in the future, a considerable proportion of general staff officers will also be efficient pilots, and qualified to undertake single-handed reconnaissance.   The

---

182. Again, I point to the U-2 and SR-71 as examples of single-seater reconnaissance aircraft with both speed and long-range. But today, the military (as well as intelligence agencies) have remote controlled aerial drones that are reported to be able to fly 400 nautical miles (460 miles or 740 km) to their targets, loiter overhead for twelve to fourteen hours gathering intelligence, and then return to base.   The entire flight is controlled by a pilot(s) tens-of-thousands of miles away.

183. Another of his predictions that has proved true.

184. This role is now performed by photographic interpreters; a unique type of intelligence analyst.

employment of trained staff officers on this duty will naturally produce more accurate and more complete information, and if commanders of troops should still be unsatisfied, it is always possible for them to reassure themselves by verifying, by personal observation, the information supplied by their subordinates.

The inventions of modern science, as adapted to warfare, have hitherto tended to loosen the control of a commander over his troops in battle; the perfection of long-range weapons has caused such dispersal of troops that even with improved methods of communication, it has been impossible for commanders to keep themselves thoroughly informed of the position and condition of their forces in action. The aeroplane has altered all this; there is now no reason why a chief commander should not have the same knowledge of the progress of an action, even if his battle line be 50 miles long, as Wellington had, and used so decisively at Salamanca. But in order that he may be able to secure this advantage, the commander of the future, as well as his staff officers, must have some experience of observation from the air. Flying, either as pilot or as passenger, has become, in fact, an essential part of the education of an officer who aspires to advancement in his profession.

1805—The Ulm Campaign

This campaign proves the disastrous effects of errors committed in concentration.

General Mack on Sept. 18 advanced to the line of the Iller and took up a position between Ulm and the Lake of Constance with advanced posts in the Black Forest. His plan was to act on the defensive, and to await the arrival of the Russians from the East. Mack miscalculated—

(a) the strength of Napoleon's Grand Army; and

(b) the marching power of the French.

He allowed himself to be deceived by the demonstrations of Murat in the Black Forest.

Aerial reconnaissance would have given Mack early information of: (1) the lines of advance of the French divisions; (2) of the screen formed by Ney's Corps to cover the movements of Lannes, Soult, Davout, and Marmont; and (3) the position of the French left.

AERIAL RECONNAISSANCE

A. Mack's Information.

Oct. 8, 1805.

B. Aerial Reconnaissance, *Oct.* 1-8, 1805.
Showing Napoleon's Lines of Advance.

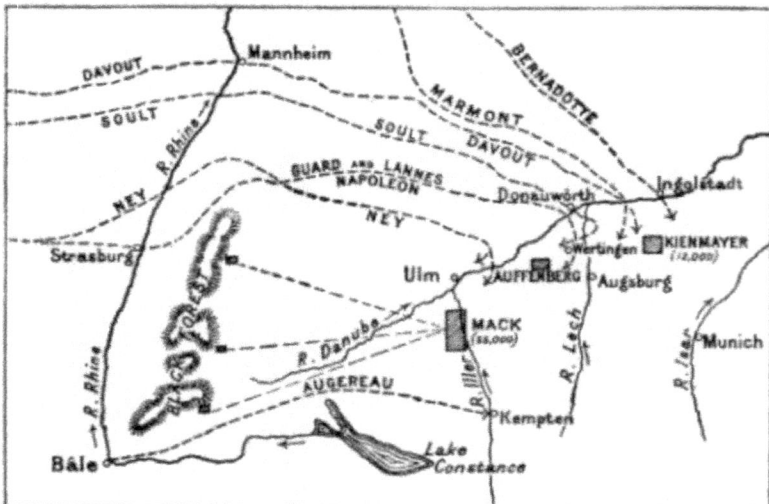

AERIAL RECONNAISSANCE

A. NAPOLEON'S INFORMATION.

Midday, June 17, 1815.

June 17.—NAPOLEON believed that BLUCHER, after the battle of LIGNY, was retiring along his line of communications towards LIEGE; the chance capture of a Prussian battery on the NAMUR Road tended to confirm this belief.

June 17.—Napoleon believed that Blucher, after the battle of Ligny, was retiring along his line of communications towards Liege; the chance capture of a Prussian battery on the Namur Road tended to confirm this belief.

AERIAL RECONNAISSANCE

B. AERIAL RECONNAISSANCE.
Midday, June 17, 1815.

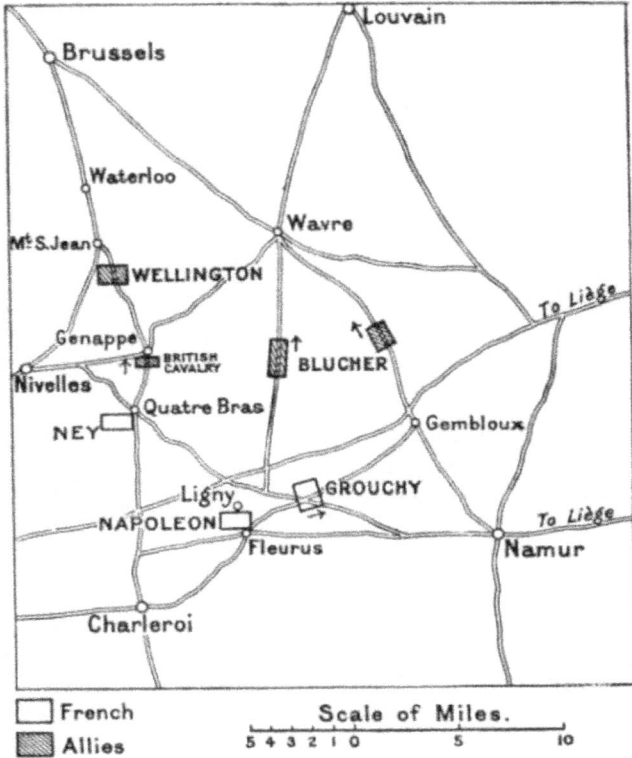

Aerial reconnaissance would have—

(1) disclosed the direction of BLUCHER's march on
WAVRE—*i.e.*, north instead of east;

(2) reported WELLINGTON's withdrawal from the
QUATRE BRAS position.

Aerial reconnaissance would have—

(1) disclosed the direction of Blucher's march on Wavre—
i.e., north instead of east; and

(2) reported Wellington's withdrawal from the Quatre Bras
position.

1849—The Campaign in Italy

Radetzky's plan was to concentrate his five corps at Pavia, interpose between the two hostile armies north and south of the Po, and then advance to Mortara against the main Sardinian army. On evacuating Milan, he fell back to S. Angiolo, and effected the desired concentration. His stratagem was successful, as the Sardinians, imagining that the Austrians were retreating, prepared to cross the Ticino opposite Magenta, and to march upon Milan. Ramorinc, under the same delusion, remained stationary to the south of Mezzano Corte.

Aeroplane reconnaissance on March 19 would have disclosed the direction of Radetzky's march.

AERIAL RECONNAISSANCE

A. SARDINIAN INFORMATION.
March 20, 1849.

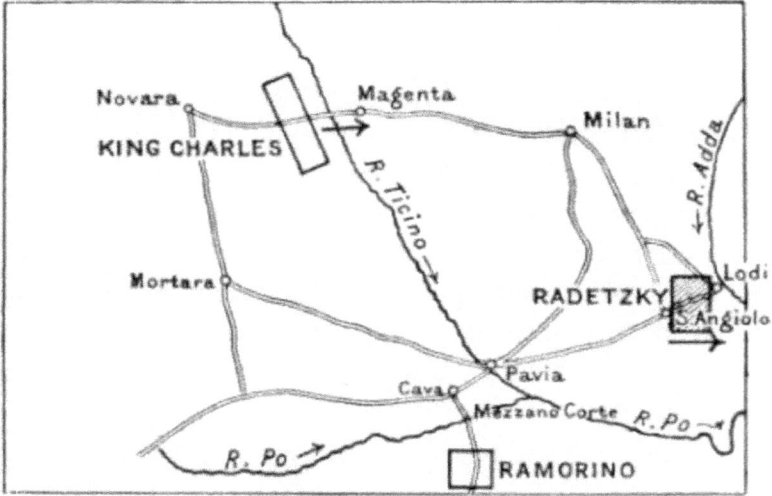

B. AEROPLANE RECONNAISSANCE.
March 20, 1849.

Austrians

Sardinians

Scale of Miles

1862

Pope, on the morning of Aug. 28, fully believed that Jackson would remain in the entrenchments at Manassas. He accordingly advanced on that position at dawn on the 28th from his camp at Bristoe Station, and directed M'Dowell [sic], whose forces lay between Gainsville and Thoroughfare Gap, to march with every man on Manassas. By this latter order Pope opened the road by which Longstreet could unite with Jackson. Aerial reconnaissance would have proved—

(a) that Jackson had retired from Manassas Junction to the old battlefield of First Bull Run;

(b) that to move M'Dowell [sic] would open the way for the union of Longstreet's and Jackson's forces; and

(c) the uselessness of Pope's subsequent order to move on Centreville.

AERIAL RECONNAISSANCE

A. POPE'S INFORMATION.
Morning, Aug. 28, 1862.

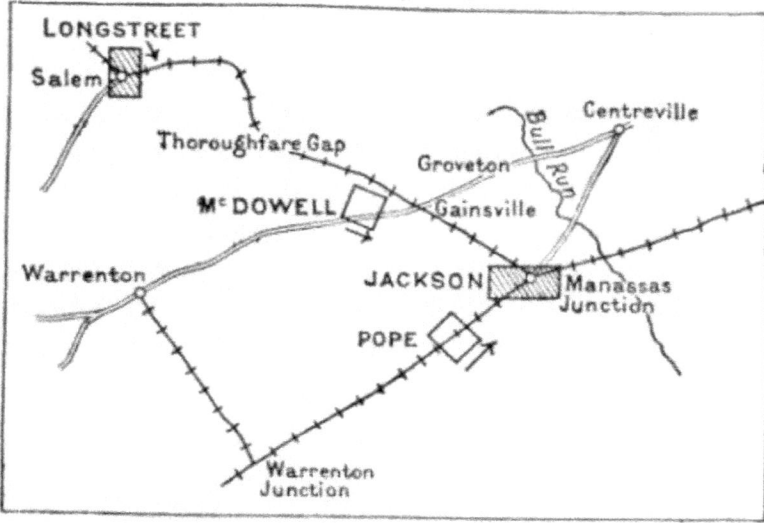

B. AERIAL RECONNAISSANCE.
Morning, Aug. 28, 1862.

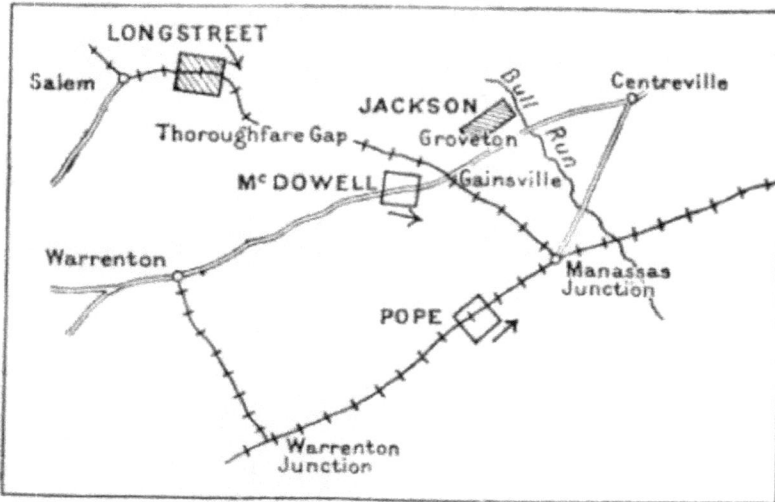

Federals
Confederates

Scale of Miles
5 4 3 2 1 0   5

HENDERSON'S ORIGINAL NOTES

A. *History of the Consulate and the Empire*, Thiers.

B. *History of the Consulate and the Empire*, Thiers.

C. *History of the Consulate and the Empire*, Thiers.

D. *The Battle of Spicheren*, G.F.R. Henderson.

E. *The Battle of Spicheren*, G.F.R. Henderson.

F. Translated by Major d'Arcy Legard, 17th Lancers.

G. Vegetius, lib. 3. *De re militari*, cap. ult.

H. *Civil War in the United States*, Wood and Edmonds.

INDEX

advanced guard, 17, 20, 44–46, 47, 56, 64, 128, 135, 139; defined, 57n77

aerial: drones, 175n182; observers, 21; observation, 22, 167, 170, 172; reconnaissance, 21, 163–176, 177, 180, 183; scouting, 5, 167–168, 172, 173

drones, 18n36, 175n182

American Civil War, 2, 31, 31n52, 115

Appomattox Court House, 31n52

Archduke Albert of Austria, 27, 28

Ashby, Turner, Jr., 36, 36n61, 80

Axarquia, 63

balloons, 21n39

Battle of Bull Run, 33n56, 183

Battle of Fort Sumter, 31n52

Battle of Manassas, 33, 33n56, 183

Battle of Omdurman, 43, 43n67

Bazaine, François Achille, 14, 14n25, 17, 18

Benningsen, General, 14–15

Black Forest, 177

Boers, 62, 93, 112

Boers, 1, 62, 93, 112

Bonaparte, Napoléon, 10n12, 11, 11n16, 13, 14, 15, 16, 17, 27, 75, 177, 179; Napoléon III, 11n16, 14

Napoléonic Wars, 13n22, 88n119

Bourbon, Louis Joseph. *See* Vendôme

cavalry, 13, 16–20, 22, 23, 52, 55, 57, 74; defined, 13n21; French, 13, 14, 26; German, 23; large scale action by, 47, 75–76; raids, 31–32, 34; roles of, 169, 170; types of forces, 23–30, 42

chain-of-command, 150n151, 150n152, 151n153, 160n161

Charles, Prince Friedrich Karl Nicolaus, 25, 25n47

Cherfils, General Pierre Joseph Maxime, 24, 24n44, 57

CIA, 12n18, 172n176

communication, 60–61, 66, 67, 67n91, 91–92, 103, 114, 116, 119, 123, 124, 125, 127, 137, 149, 152–153, 155, 158, 176, 179; blocking, 49, 66; drums, 50n71; flags, 50n71; radio (wireless), 50n71, 124n140; raids on, 33 34; runners, 50n71; signals while on patrol, 124–125, 124n140; telegraph, 50, 76; telephone, 76, 76n105; verbal, 123

(autonomous), 20, 20n 37, 21; ground, 142–149; protective, 20, 20n37; contact, 20, 20n37; night, 88–94; risk assessment, 101; satellites, 172n175; tracking enemy, 102, 109–110; enemy estimation, 111–112

reports. clearness and brevity, 161; maps, 145; self-validating, 151n155; writing, 82–83, 82n115, 147; verbal, 147

Rodrigo, Ciudad, 108, 108n130

Russian Empire, 34n57

Russo-Japanese War, 34n57

Saar, 16, 16n32

sabotage, 80n111

Salamanc, Battle of, 10, 10n13, 108n130, 176

scouts, 76–115, 116–119, 122–124, 125–127, 160–161; aerial, 167–168, 172; enemy, 41, 49; French, 18; German, 37; Prussian, 16

security, defined, 40

Sédan, Battle of, 10, 10n15

Seydlitz, Friedrich Wilhelm Freiherr, 68, 68n94

Shenandoah Valley, 36

social engineering: misuse of term, 52n75

Soudan, 65

South African War, 62

Soviet Union, 166n170

Spicheren, Battle of, 13, 13n23; map, 17

SR-71, 165n169, 167n171, 175n182. *See also* U-2, Powers, Francis Gary

Stanhope, Philip Henry, 64n85

Steinmetz, General Karl Friedrich, 25, 25n48

Stuart, James Ewell Brown (Jeb), 32, 32n53

Sudan, 65n88

Sun Tzu, 1, 35n60

telescope, 101, 101n127

Tirah, Pakistan, 65, 65n89, 65n90

topographist, 118, 118n138

training. *See* education

traps, avoiding, 103–104

U-2, 166n170, 167n171, 172n176, 175n182

Ulm, Battle of, 10, 10n12,

Vendôme, Duke of, 64, 64n86

War of Spanish Succession, 63, 63n82

Wellington, Duke of, 10n13, 73, 108, 108n130, 109, 172, 176, 180

World War I. *See* Great War

www.ingramcontent.com/pod-product-compliance
Lightning Source LLC
Chambersburg PA
CBHW031131090426
42738CB00008B/1050